REVOLUTIONARY IRAN AND
THE UNITED STATES

US Foreign Policy and Conflict in the Islamic World

Series Editor
Tom Lansford
University of Southern Mississippi, USA

The proliferation of an anti-US ideology among radicalized Islamic groups has emerged as one of the most significant security concerns for the United States and contemporary global relations in the wake of the end of the Cold War. The terrorist attacks of September 11, 2001 demonstrated the danger posed by Islamic extremists to US domestic and foreign interests. Through a wealth of case studies this new series examines the role that US foreign policy has played in exacerbating or ameliorating hostilities among and within Muslim nations as a means of exploring the rise in tension between some Islamic groups and the West. The series provides an interdisciplinary framework of analysis which, transcending traditional, narrow modes of inquiry, permits a comprehensive examination of US foreign policy in the context of the Islamic world.

Other titles in the series

Revolutionary Iran and the United States
Low-intensity Conflict in the Persian Gulf

JOSEPH J. ST. MARIE
University of Southern Mississippi, USA

SHAHDAD NAGHSHPOUR
University of Southern Mississippi, USA

Routledge
Taylor & Francis Group

LONDON AND NEW YORK

First published 2011 by Ashgate Publishing

Published 2016 by Routledge
2 Park Square, Milton Park, Abingdon, Oxon OX14 4RN
605 Third Avenue, New York, NY 10017

Routledge is an imprint of the Taylor & Francis Group, an informa business

British Library Cataloguing in Publication Data
St Marie, Joseph J.
    Revolutionary Iran and the United States : Low-intensity Conflict in the Persian Gulf.
    – (US Foreign Policy and Conflict in the Islamic World)
    1. Iran – Foreign relations – United States. 2. United States – Foreign relations – Iran.
    3. Iran – Foreign economic relations – United States. 4. United States – Foreign economic relations – Iran. 5. Iran – History – Revolution, 1979–Influence.
    6. Low-intensity conflicts (Military science) – Persian Gulf Region – History.
    I. Title II. Series III. Naghshpour, Shahdad.
    327.5'5073-dc22

Library of Congress Cataloging-in-Publication Data
St. Marie, Joseph J.
    Revolutionary Iran and the United States: Low-intensity Conflict in the Persian Gulf /
    by Joseph J. St. Marie and Shahdad Naghshpour.
        p. cm. – (US Foreign Policy and Conflict in the Islamic World)
    Includes index.
    1. United States – Foreign relations – Iran. 2. Iran – Foreign relations – United States.
    3. Low intensity conflicts (Military science) – Persian Gulf. I. Naghshpour, Shahdad.
    II. Title.
    JZ1480.A57I718 2011
    327.73055–dc22   2010048958

ISBN 13: 978-0-7546-7670-6 (hbk)

We dedicate this work to the memory of our parents:
John H. & Josephine St. Marie
Ali & Azam Naghshpour

Contents

Contents

Acknowledgements

Great thanks go two of our graduate students Mr. Rian Plaswirth and Mr. Jim Kierulff who provided helpful assistance.

Chapter 1
Low-Intensity Conflict and Fourth Generation Warfare

The relationship between the Islamic Republic of Iran and the United States is for all practical purposes a non-relationship with each nation pursuing adversarial and hostile policies and actions towards the other. The reality of the situation is that Iran and the United States are nations engaged in a low-intensity conflict. This work examines the low-intensity conflict between the United States and Iran, with respect to the military, economic, and political aspects in which the conflict has taken in the past thirty years.

This chapter investigates the relationship between The Islamic Republic of Iran and the United States of America before and after the 1979 Iranian Islamic Revolution with particular emphasis on the post-revolutionary war stages. The analysis examines the adversarial relationship, fourth generation warfare (political, economic and diplomatic war in the 21st. Century), and the theoretical constructs that surround this form of warfare between states. Furthermore, this chapter sets the stage for subsequent chapters by demonstrating how domestic politics and low-intensity warfare are linked. This linkage, we argue, determines the ebb and flow of the conflict.

The Adversarial Relationship

The effective date of this conflict can be pinpointed as 4 November 1979, when Iranian radical students seized the US embassy in Tehran, taking US diplomats and civilian workers hostage for 444 days. This action, which was tacitly sanctioned by Ayatollah Ruhollah Khomeini, produced a long lasting and significant change in US foreign policy toward Iran (Jordet nd, Pollack 2004, Quosh 2007). The diplomatic crisis was highlighted by the failed American rescue attempt of the hostages, which put each nation firmly in a confrontational state. Subsequent confrontations were manifested through moral or ideological struggles between two opposing ideologies pursued by means other than direct military confrontation.

Similar to Western political practice, the Islamic Republic has democratic institutions and constitutional checks and balances. However, these governmental characteristics are derived from Islamic law which is inherently undemocratic. For example, candidates for political office are required to submit their credentials to religious scholars before they are certified to stand for election (Abrahamian 2008, Thaler et al. 2010). The Constitution of the Islamic Republic is derived from Shia Islamic law, which in turn is derived from the Koran, the "infallible word of God." The Islamic Republic has created a political system diametrically opposed to the American and Jeffersonian ideals of "life, liberty, and the pursuit of happiness" through the implementation of various social, economic, and political controls. The Iranian government sees the American constitution and political system as corrupt and devoid of the perquisite religious component, which they deem necessary for a just society. With such diverse ideological views there has been little common ground for compromise or co-existence. Those episodes of cooperation such as "Arms for Hostages" have tended to end badly. Even Iranian assistance in the defeat of the Taliban was short lived after Iran was made part of the "Axis of Evil" short of direct military conflict both have sought various means to demonstrate their ideological superiority within political and economic arenas.

Levels of Analysis

To begin, a meaningful comparison of Iranian and American interaction must examine the levels of analysis to determine how they relate to low-intensity conflict and fourth generation warfare. To be sure, it is generally accepted that individual, state, and systemic or global levels of analysis are proper domains of inquiry for the study of international relations and, more specifically, conflicts. The present analysis crosses various domains as an essential function of the nature of modern conflict. Since modern conflict is multi-dimensional our analysis will be multi-dimensional.

The individual level of analysis is primarily concerned with the motivations, constraints, and incentives a decision maker works under. While there are incentives and constraints imposed upon all decision makers, the personal traits of a policy maker are appropriate for the study of low-intensity conflict and fourth generation warfare (Lind 2004). Given the overarching ideological or moral component of this new type of warfare, personal attributes become important when examining individual and collective motivations or preferences for certain policy actions.

Looking at the state and its internal attributes to determine policy direction is another part of fourth generation warfare analysis. Clearly the alignment, relative power, and ideology of constituent groups within a state have a profound impact on the policies a state may pursue. The fragility of the ruling coalition, strength of the opposition as well as economic and military threats all impact decisions made by elites within the state. Including the state level of analysis in our conflict model is appropriate as specific tactics and strategies are constrained by domestic factors as well as personal and systemic factors (Putnam 1988).

The global or systemic level of analysis emphasizes the impact decision makers have on the global political economy. Given that fourth generation warfare includes economic elements, the inclusion of economics within the global level of analysis is reasonable. In addition, such analysis allows for the interaction of global markets with domestic factors and the impact on the decision making process. While we do not examine conflict that is global in respect to the number and location of the participants we do examine conflict that has global implications. For example, a blockade of the Straits of Hormuz would have a profound economic impact on industrialized countries. Conversely, lack of revenue would cause serious social and economic disruptions in the Persian Gulf States (Barnett 2005, Cordesman 2006, 2009, Hugill 1998, Krepinevitch et al. 2003, Larson et al. 2004, North & Choucri 1983, Truver 2008). While it is generally considered a taboo to mix the levels of analysis, in this work it is essential to understand the complicated nature of fourth generation warfare (see Waltz 1979 for a discussion of levels of analysis). The nature of low-level conflict is necessarily one that requires an analysis of those decision makers who wield power, the domestic sources of policy, and how such policies will affect the larger international system. One aim of this study is to validate the concept of fourth generation warfare as a form of total warfare that encompasses all aspects of decision making.

Low-Intensity Conflict and the Fourth Generation of Warfare

Low Intensity Conflict

Low intensity conflict has been characterized as a war that never reaches a critical mass of men, material, and the will to definitively win. It is also a tool employed by weaker states, groups, and movements that seek some sort of concessions. Low intensity conflict differs to a great extent in so far as it may be prolonged, it may ebb and flow, and it may have various levels of intensity, yet resolution is elusive (Liang & Xiansui 1999).

Fourth Generation Warfare: What it is and What it is Not

The term fourth generation warfare has only been in the security lexicon since 1989 with the publishing of an article in the Marine Corps Gazette (Lind et al. 1989). In this article, fourth generation warfare is seen as an evolution of warfare whereby the first generation is characterized by muskets and static lines firing upon each other at short ranges. The second generation of warfare is characterized by rifled guns, machine guns, indirect artillery fire, and movement as opposed to static tactics. The third generation of warfare begins in 1918 and focuses on maneuver warfare where flexibility and bringing dispersed units together quickly to overwhelm an adversary brings victory. Each of these evolutions has produced increased casualties, material damage, and economic costs to the adversarial countries. The fourth generation can be described as 'networked' or 'technology' driven warfare. Technology has driven many of the advances since the culmination of the air-land battle of the third generation (the Gulf War of 1991). Thomas X. Hammes sees fourth generation warfare as both a strategic and tactical endeavor (Hammes 2005, 2007).

The strategic side of modern fourth generation warfare is reversed from previous forms of warfare. For example, a fourth generation war need not have territorial ambitions, nor will the linear battlefield be the primary ground where battles will be fought. Conversely, opponents attempt to change the minds of enemy decision makers through various means of persuasion utilizing the tools of fourth generation warfare. These means are what make fourth generation warfare distinct from previous forms of warfare. The ways and means of this new type of warfare can be summed up simply as "total war" meaning that any type of weapon be it cyber-attacks, conventional military attacks, public relations smears, or economic undermining can be considered a weapon in the arsenal of fourth generation warfare. Thus, for purposes of this work, we define fourth generation warfare as: *conflict between two or more nation-states aimed at persuading/ coercing the other to bend to its demands, whether they be military, economic, or political; whereas the means of conflict are not restricted to military or uniformed combatants; whereas political, military and economic means are used as weapons to gain persuasive advantage.* Fourth generation warfare is thus "total" warfare where the only rule is that there are no rules (Liang & Xiansui 1999).

Characteristics of Fourth Generation Warfare

Fourth generation warfare may be characterized as more of a moral struggle rather than a set-piece or maneuver battle. Moreover, this type of warfare is predicated on the examination of various political, economic, and military networks of an adversary and attacking at specific nodes that will cause widespread damage within and between networks. For example, a fourth generation warfare attack on Pakistan by India to "persuade or compel" Pakistan to refrain from supporting terrorist organizations intent on attacking India (Bombay hotel attack type of operation) might include a cyber attack on military and economic targets, such as the air defense network and the power grid. A hypothetical example of a political attack might involve jamming of Pakistani television and radio stations and inserting propaganda or disinformation. Accompanied by small and limited conventional military incursions, insurgent infiltrations, and small scale attacks, India could "persuade" Pakistan to change its policy, if sufficient "damage" was inflicted. Fourth generation warfare takes advantage of the complexities in military infrastructure, economic systems, and political approval to achieve its goals. The importance of examining conflict through the lens of fourth generation warfare is that the boundary lines between civilian and combatant; between permissible targets and off-limits targets; between economic and military targets; and differences between political systems are all blurred. Modern warfare is not the black and white of the old days but a continuum from black to white with fourth generation warfare comprising a large gray area between each pole (Liang & Xiansui 1999).

Fourth generation warfare is a type of warfare that seeks to undermine the unique and specialized strengths an enemy might have. For example, in Operation

Iraqi Freedom and its aftermath insurgents faced a much stronger opponent yet were able to destroy or disable the American M-1 Abrams main battle tank, an armored vehicle that has few peers on the battlefield. In a similar way, roadside bombs, kidnappings, intimidation, and demonstrations all undermined the feeling that the American military would be able to protect the population. To undermine the enemy's strength highlights vulnerabilities which lead to a loss of confidence by the population of the defender. Four-generation warfare is not insurgency since it is played out in the diplomatic and realm of states as opposed to intrastate violence. Loss of confidence is one key aspect of how fourth generation warfare attempts to persuade a government to change its policies. In this case, a democratically elected government may find itself faced with significant domestic unrest as a result of being attacked, and the perception of citizens that it is defenseless in the face of renewed attacks. Undermining the confidence in the population is one strategy an aggressor can use to create advantages even when it is militarily weaker. Authoritarian governments are even vulnerable to unrest as the result of lost battles with weaker adversaries such as with Russia in 1905. Strongmen rely upon a system of redistribution of rents to hold on to power. If this system is interrupted their hold on power is reduced and less absolute and they become susceptible to coups or invasion. Modern warfare is predicated upon exploiting the enemy's weaknesses on the battlefield. However, fourth generation warfare widens the battlefield to include politics and economics; thus, the numbers of avenues of attack are increased exponentially. Militarily, weaknesses may be found in persons who lack training, in weapons systems that are non-operational or too complex for conscripts to operate. The military leadership is another aspect that may be attacked either directly (in the case of a diversionary battle) or individually (through assassination). Any aspect of the military is a target such as personnel, leadership, or equipment.

Recent history has shown that complex weapons systems can be rendered useless when faced with targets for which they were not designed. Examples include the M-1 main battle tank and American fighter aircraft in Iraq and Afghanistan. The lightly armored humvee was found to be highly susceptible to small roadside bombs, something the designers never contemplated. In some cases highly complex weapons do not function well when confronted with simple weapons. Fourth generation warfare sees persuasion to change policies as a multi-faceted problem bringing to bear any and all weapons at an adversary's disposal to render so-called strengths weak and to exploit these weaknesses.

Another characteristic of fourth generation warfare is the presence of differential weaponry and operations. Related to the notion of exploiting an enemy's weaknesses, a weaker opponent uses non-standard and improvised means to attack a stronger opponent, through economic, political or military means. For example, in Iraq, demonstrations are lead by disaffected groups centered on religious festivals and holy shrines. Any attempt by American troops to break up the demonstration would be met with hostility based upon religious and cultural beliefs. Such use of culture and religion by adversarial groups is an excellent example of the sort of differential tactics used against a stronger force. Differential power levels and abilities are characteristic of fourth generation warfare that leads each side to change its tactics in response to moves by the other side. It is not important which side moves first. The other is obligated to follow suit. In conventional warfare, reacting to an enemy is regarded as allowing the enemy to dictate terms of the battle, such as choosing the site, weapons, *etc.* However, in fourth generation warfare, dictating the ebb and flow of the overall conflict may not be possible since changing tactics in the political, military, and economic spheres would require a high level of coordination and sophistication (Arquilla & Ronfeldt 1999, 2000, Dorman 2008, Edwards 2005, Hoffman 2009, Liang & Xiansui 1999, Singer 2002, Tangredi 2000).

The global or systemic factors that drive fourth generation war can be seen at all levels of analysis. The nation-state is the building block of the international system and will continue to be so for the foreseeable future. However, the forces of globalization, information technology, and international migration have lessened the sovereignty of the state in some areas. It is precisely these regions where fourth generation warfare finds its impetus and mode of operation. The state in many ways has lost its monopoly on the use of force. If an individual or group faces persecution in one state, moving to another is a relatively simple task.

Modernization has placed the state in a precarious position versus its citizens. While having the means to place surveillance on individuals, civil rights can be violated, and cause must be shown for spying on individuals or groups. Clearly, the power of the state is limited, ironically, by the state's own judicial system. The loss of sovereignty by the state is not just the realm of democracies; authoritarian regimes can also lose control over their population in a different manner. The democracy will find its control over society constrained by its legal system as mentioned above. The same rights used to protect citizens from the state will be used to limit the state's power in rooting out subversive activity or to preempt hostile groups from attacking state edifices. The authoritarian state will be able to repress and use surveillance at will, yet will not be able to quell underlying dissatisfaction and rebellious groups. Sri Lankan military forces have faced stiff resistance by the Liberation Tigers of Tamil Eelam since 1975 and have not fully destroyed the organization, although most hostilities ended in 2009 with the defeat of the Tamil Tigers and death of their leader (Edwards 2005, Ganguly 1996, Hammes 2007, Kinross 2004, Truver 2008). Even serious curtailment of civil rights and violations of human rights were not enough to break the power of the group.

Simply put, the power of the state cannot destroy an idea. Democracies can co-opt groups within a legal framework, while authoritarian systems can limit the potential to undermine the regime at best. Fourth generation warfare makes use of the internal vulnerabilities of the government to develop its attack plans and attempts to use the vulnerabilities of the state as a form of persuasion.

The range of conflicts in the modern state system gives rise to a second factor that makes fourth generation warfare more applicable today. Since the end of the Cold War cultural, ethnic, and religious conflicts have been rampant. Some are based upon historical grievances; others are for economic gain by rival ethnic groups. Yet, religious conflicts have taken place in other countries (East Timor is an example) usually over land or some religious site. Fourth generation warfare is a product of the demise of the old Cold War global system and the new types of conflict that has occurred since. Each type of conflict tends to be particularly bloody, in human terms, which limits the ability of the warring sides to reach a settlement. In many instances partition is the best solution as exemplified in East Timor where a new nation emerged from the conflict. Cultural, ethnic, and religious conflicts tend to occur more often when a nation is perceived by dissidents to have lost some of its sovereignty (Curtis 2005, Flanagan & Schear 2007, Hammes 2005, Hoffman 2005, Liang & Xiangsui 1999).

Finally, globalization has been a contributing factor toward fourth generation war as well as a catalyst for this new type of warfare. Globalization can be seen as economic, cultural, and political. Economic globalization revolves around capital and trade flows. Increased trade is a way to obtain foreign currency as well as bring prosperity to the population. The darker side is that in many cases few actually benefit from globalization, as wealth tends to become more concentrated in many instances. Moreover, imports can inflame long held cultural and religious hostilities. Likewise, free movement of capital cause severe economic strains if portfolio investment is suddenly withdrawn (Friedman 2000). Cultural globalization includes those factors that spread or attempt to create one global culture. Access to media, in particular satellite television, has made news instantly available to far more people than newspaper or radio could ever do. Combined with dramatic color pictures, television has the ability to polarize sub-national groups or even entire nations in a relatively short time. Finally, political globalization is the spread of diplomatic and non-governmental organizations to new states and the intensification of ties with states that already have significant contact. On the whole, various forms of globalization can and do contribute to an environment that is conducive to low-intensity or fourth generation warfare.

Fourth Generation War (FGW) Theory

The evolution of FGW began as nation-states became central actors in the global system in the sixteenth century. Previously, religious organizations, social movements, and tribal groups held sway and inhibited the state from exercising complete sovereignty over its territory. As the central actor in modern international politics, the state is at the center of FGW as in previous eras of warfare. In as much as the state is the central actor in modern international politics and in past eras of warfare, it is also at the center of FGW. What makes FGW substantively different is its all-encompassing focus on warfare as opposed to Clausewitz's notion of 'politics through military means.' Two primary movers have influenced FGW in the past twenty years. These are Lind et al. and Hammes along with cogent commentary by Curtis and Echevarria.

The Lind Perspective and FGW

Lind asserts that warfare has evolved in eras where innovation in technology and tactics has caused profound changes in the way militaries and policy makers approach warfare. For instance, defense and offense have highly different mechanisms depending upon the era of warfare. Offense in siege warfare bears little resemblance to modern land-air battle from both the offensive and defensive perspectives. However, the enlarged battlefield found in FGW encompasses outer space and cyber-space, two domains not previously encountered by policy makers and war fighters (Lind nd, Lind et al. 1989).

Another element of the Lind perspective on FGW is the notion of various eras of warfare. While criticized by some, this perspective neatly categorizes the history of warfare into four eras (Curtis 2005). A limited version of this breakdown is presented. The first era, or first generation of warfare, was post-1648 until the First World War. First generation warfare was, for the most part, conducted by set-piece battles where the objective was almost always political domination. The victors would, in essence, take as their prize the defeated state. This type of warfare was straightforward—defeat the enemy entirely with military force.

In second generation warfare Lind et al. (1989) describes what is found in World War I French Military doctrine of the day. Punctuated by massed artillery fire, troops would move toward the enemy in infantry assaults that resulted in high casualty rates. Aircraft were also used to provide some sort of support for attacking infantry. We should however note that the German military developed successful tactics that neutralized the French tactical innovations, as described by Lupfer (1981). One tactic, in particular, that tended to nullify allied advances was a counter attack initiated before the allies could consolidate their gains. In short, World War I introduced new command and control features to the battlefield that made the control of ground units more manageable and well integrated with artillery and aerial reconnaissance into the overall battlefield.

Third generation warfare encompasses World War Two and subsequent battles of maneuver culminating in Operation Desert Storm, the liberation of Kuwait, and the defeat of Saddam Hussein's military. The striking features of this era are the integration of artillery fire-support and maneuver into the battlefield. Integration of close air support was perfected by Wolfram von Richtofen in the beginning stages of World War II. It was further refined in Korea, Vietnam, and finally in Iraq. The focus on maneuver, commonly referred to as the Blitzkrieg, was the initial innovation that culminated in the American doctrine of the air-land battle as implemented in Iraq in 1991. Maneuver by larger and larger units attempting to penetrate, destroy, or out flank fixed or maneuverable units of an enemy is the hallmark of this era of warfare. The logic of the battlefield was the same for Rommel, Patton, Sharon, and Schwarzkopf. However, this logic began to change with the invasion of Iraq or Operation Iraqi Freedom. The integration of large units was relaxed in favor of smaller more mobile units which could quickly converge, form a larger striking force and defeat an enemy, then disperse and move only to repeat the process as the need arose. This sort of quickly evolving type of warfare was made possible by increasingly sophisticated electronics which boosted the lethality of not only the common soldier but of the munitions used by ground, air, and naval forces. Increasing accuracy and lethality, enabled by electronic networking, allowed American forces to defeat Iraqi forces in a minimum of time with minimal troops (Arquilla & Rofeldt 2000, Dorman 2008, Dunlap 2006, Lian & Xiangsui 1999, Lock-Pullan 2005, Metz 2000, Watts 2004).

The interregnum between the fall of Saddam Hussein and the Iraqi insurgency can be seen as a time of transition from third to fourth generation warfare. For the Lind group of authors we can surmise that the Iraqi insurgency and the war in Afghanistan are the start of fourth generation warfare. The question remains, what is the Lind et al.'s concept of FGW and how has that played out in contemporary history? We can only speculate as to what Lind et al. may be getting at when they describe FGW. Given their overall theoretical direction they seem to indicate a renewed focus on insurgency, but an insurgency that takes the classic notion one step further by integrating the idea of increasing use of technology in FGW.

One criticism of Lind *et al.* has been leveled by Curtis who compares the thought of Trevor Dupuy (1990) with that of the Lind group (Curtis 2005, Dupuy 1990). They have somewhat differing notions of innovation and war, yet this is not surprising given the different directions their analyses take. The Curtis critique chides Lind for creating a dogmatic and concrete classification when he feels a more fluid system such as that presented by Dupuy is more in tune with history. This critique is valid in some ways, however Dupuy speaks in micro terms of the technology employed by individuals and the interaction of individual level technology and macro level tactics (Dupuy 1990). The Lind group, on the other hand, views the history of warfare in macro level terms and emphasizes major shifts such as armored maneuver warfare versus set piece infantry battles (Lind nd, Lind et al. 1989). The importance of the Dupuy study is the linkage between technology and tactical innovation. This is not mutually exclusive with Lind despite the analysis Curtis provides. Dupuy and Lind are looking at the same phenomena but through a different lens—an idea not covered in critiques. The Lind group seems to be approaching FGW as a Kuhn paradigm shift. However, this is not the case. A close reading reveals that the generational approach to warfare is not intended to be an analysis of a paradigm shift but in fact a more humble analysis of how policy makers should approach technical and strategic innovation and how these innovations are carried out on the battlefield. The present study argues that the Iran-US low intensity conflict has all the characteristics of the FGW.

The importance of FGW for this study is in the theoretical stance it takes on the interactive forces of politics, military power, and economics. Each component is used to gain advantage over the adversary and sometimes they are used in combination (Liang & Xiangsui 1999). Currently, war is seen as more of a total war. When fought at a low level of intensity each side uses each of its forces to achieve dominance. Future chapters demonstrate how the low-level of conflict between the United States and the Islamic Republic of Iran follows the theory of fourth generation warfare.

Hammes and Fourth Generation Warfare

Thomas Hammes provides further insight into the idea of fourth generation warfare. Hammes defines his conception of FGW as primarily a political struggle carried out by many different means and tactics (Hammes 2005, 2007). The idea of a non-linear battlefield is a common element in the Lind group and Hammes models.

While accepting the terminology of fourth generation warfare Hammes rejects the idea of generations of warfare that is a central theme of Lind et al. Rather than fixing a point where warfare underwent a paradigm shift as Lind et al. does, Hammes discards the generational notion of epics of warfare as he asserts that precursors to FGW are necessarily prior to the present situation and were, according to Hammes, enunciated by Mao in the Chinese Revolution (Hammes 2005, 2007). As a precursor to the fourth generation of warfare, Hammes identifies Mao as the innovator in warfare that initiated the transformation into fourth generation warfare. Rather than providing a paradigm shift as the Lind group has done, Hammes sees the same linear advancement in stages of warfare but is less dogmatic in his approach. Thus, while the Lind group sees paradigm shifts or a staircase of learning, Hammes sees a slow, steady progression where innovation in one conflict may not influence the next conflict in historical linear order. Rather, innovation proceeds by osmosis throughout military history with innovations perhaps only being fully realized many years later. The role of tanks in WW I, WW II, and Desert Storm is an example. The firepower, mobility and speed of the tank evolved yet the tactic did not undergo paradigmatic changes given that the basic function of the tank remains the same.

Historically we can see Mao turning orthodox Marxist doctrine on its head by basing his revolution—partly out of necessity and partly out of theory—on the peasantry. Mao's innovative way of fighting the Japanese consisted of guerrilla raids and insurgency. Being able to create cells that were quickly brought together for action and as quickly dissolved gave the Communists a great advantage over the less agile Japanese (Fairbanks 1986, Hammes 2005, Mao 2009). The Communists correctly reasoned that Japanese reprisals would be a major factor in recruitment, tipping the ratio of peasants who were revolutionary versus the status quo. Correctly surmising that citizens would be offended by reprisals (even though they may not be actual communists or revolutionaries) helped Mao build a willing pool of either active insurgents or material supporters for the insurgency. Thus, Mao utilizes intrinsic and realistic methods to solve an ideological and military problem. Such innovation is at the core of the Hammes and the overall theory of FGW.

Hammes deviates from the Lind group in his reliance on signaling, the signaling of intentions, to ground his conception of FGW or this new form of total war. In the simplest case, one adversary signals its intention to the other via some channels. These channels need not be military but can also be political (both diplomatic and alliances) and economic (sanctions or direct threats to disable an adversary's economy). Political signaling can be through mutual diplomatic means or third parties. Alliances can be made or reinforced through some sort of enhanced economic or military cooperation. Many tools are available to policy makers in the political arsenal. Politically, a nation may be encouraged to isolate a target nation with certain incentives such as diplomatic treaties, economic rewards, or even promises of protection for special rights like basing of military forces. Numerous economic tools abound, including sanctions, restrictions on the use of the global financial infrastructure, and even attempts to limit exports or imports via interdiction or through the use of bureaucratic 'red tape.' Military signaling can take many forms as well, from increased surveillance to direct violation of a target nation's sovereignty by ships, aircraft, or troops.

Signaling can be a complex undertaking. Signaling can have unintended consequences or the message may not be interpreted in the correct manner. For example, the book (Jervis 1970, 1976; Moore & Galloway 1992), and subsequent movie "We Were Soldiers", depicts the battle of Ia Drang in 1965. The battle was perceived by policy makers as a tactical victory for the United States where US mobility and firepower inflicted such sufficient causalities on North Vietnamese forces that in a war of attrition the North Vietnamese would eventually capitulate. This, however, was not the case. The message of "we will wear you down" was well received by the North Vietnamese. However, while acknowledging this message it fell upon deaf ears as the North Vietnamese were fighting a war of liberation; they would endure severe causalities to conquer the whole of Vietnam since defeat was not an option. Clearly, signaling relies upon not only effective communication of the intent but the message must be tailored to the recipient. The message must be one that is clear to the adversary and within their frame of reference for the conflict or potential conflict.

Another example from the same era is the Rolling Thunder aerial bombardment campaign. Fashioned along the lines of the economic theory of Thomas Schelling, the intent was to signal the determination of the United States by increasing the pain inflicted upon the North Vietnamese. The simple plan was to start bombing above the Demilitarized Zone (DMZ) starting with small raids on relatively unimportant targets. As the raids moved north they would increase in size, intensity, and target value. By increasing the pain policy makers believed the North Vietnamese would note the US commitment to South Vietnam and cease, or at least reduce hostile activities. While this exercise in signaling was certainly clearer in intent, it too was not successful since the war aims of the combatants were sufficiently different. The message was ignored by North Vietnam.

The key to Hammes and his notion of FGW is that of signaling and the ability of both the signaler and the receiver to understand the signal (Hammes 2005). Differences in cultural, ideological, and war aims are potential interferences to effective communication. To this end Hammes envisions a four-step signaling process where, in the first step, the initiator determines what should be the message to the adversary.

Second, the sender chooses the best network for delivery of the message. There are three primary networks. The first network is the political network which includes diplomatic contact, media pronouncements, and the speeches of major policy makers. The economic network can signal through personal contacts, bureaucratic rule making, direct sanction, or limitations on trade. Third, military networks may be used for signaling one's intent. This sort of messaging is much more direct so there is no ambiguity that a message has been sent but the intent of the message can be misinterpreted. For example, the message that a country should halt the activity of arming a third party radical group may be sent by airstrikes on the groups' munitions caches. Such a signal degrades the group's offensive capability yet does not directly attack the adversary thereby increasing tensions and creating an act of war or a justification for retaliation.

Third, a signaler may attack with surgical strikes intended to destroy the weapons intended for the radical third party group within the adversary's territory. This sort of attack sends a much stronger message and is an act of war, thus care must be taken when using military force as a signal. In the former example force is applied but in a third country against the group the recipient is supporting. This use of force is direct, focused, quick, and is readily intelligible without engaging in an act of war or raising nationalistic fervor. Separating signals that are focused on policy decisions may not be popular among the adversary's electorate or the ruling coalition if there is no electorate among the faction. When using military force as a signal it is of utmost importance that the sender sends the proper message and that the message is received in the proper way. Unlike signaling used in the Vietnam War, as discussed above, the specific network that sends the signal must convey the message clearly.

Fourth, the signaler must determine if the message is received. In the case of direct military action signal reception is rapid, while in the case of political or economic messages the lag time can be considerably longer. The time for diplomatic channels to operate can be lengthy for the most part. When signaling is taking place direct signaling from one decision maker to the other is possible yet, depending on the issue and salience to each party, the response might be delayed. The importance of sending and receiving the correct signal is the heart of the theory presented by Hammes. Like two-way radios, both parties must be on the same frequency to send and receive. If both parties are not, false messages can be obtained and poor policy decisions will be made. Nations must determine an adversary's war aims to be sure that the signaling is effective and rigorous, if not the signaler will not understand the adversary and perhaps arrive at a strategy that is inherently self-defeating (Hammes 2005).

Signaling is naturally an important aspect of conflict and communicating intentions if one is ready to escalate conflict or to back down. War objectives, on the other hand, can be more than defeating the enemy. These objectives can take into account the price one is willing to pay for victory or how much one is willing to risk for victory. In either instance the differences in magnitude between war objectives of adversaries is the important notion. For example, lacking the will or resources to attack and defeat the Islamic Republic of Iran the USA can fight a low level conflict on terms advantageous to its economy, diplomatic stance, and military forces. Conversely, Iran can choose from a different set of resources and tactics or alternatively choose to fight in a manner consistent with its intention of continued sovereignty, political stability, and increasing regional influence. Fighting a fourth generation war is a logical policy if the war aims of the two combatants are such that they will avoid direct military confrontation and escalation while using proxies for direct actions against each other or their allies. Iranian support of Hezbollah in Lebanon against Israel is an example of such a linkage supporting low-level warfare (Cordesman 2006, 2007, 2009, Ganji 2006, Gordon 1998, Hajjar 2002, Smith 2007).

It is clear in this case that the low-level war aims of Iran and the US have been relatively stable and have been relatively low risk since 1979. The United Sates seeks several objectives: 1) regime change in Iran, yet short of regime change at least containment of an expansionist Iran; 2) reduce the ability of Iran to control the Persian Gulf region through military, economic, and political means; and 3) reduce the ability of Iran to support terrorist organizations and organizations that are hostile to American allies (for example, Hezbollah in Lebanon and Yemeni radicals who attack Saudi Arabia). Iran seeks: 1) to limit American influence in the region; 2) spread the ideas of the Islamic Revolution to neighboring nations, particularly nations with large Shia communities; 3) confront the US in the Persian Gulf and internationally through low-level military actions and diplomatic means; 4) engage the greater Islamic world in dialogue detrimental to US interests; and 5) create bilateral economic, diplomatic, and military ties with states hostile to US e.g. Venezuela and China. These preferences, or war aims, are in tune with fourth generation warfare and the historical record of the two combatants we address in this work.

The Curtis Critique of FGW

Curtis critiques FGW as a dogmatic system that has succeeded the traditional air-land battle with a philosophical system of supposedly irrefutable premises. While Lind *et al.* and Hammes believe in FGW, Curtis states, as Clausewitz said, "war is politics through military means." While not rejecting this definition, FGW theorists acknowledge that FGW is also inherently political and cultural. Thus, a new sort of warfare has emerged similar to the Chinese notion of "Total War" (Lian & Xiangsui 1999). This notion does not encompass knee jerk "Mutual Assured Destruction" but a more nuanced notion of war through various means. Such various means are at the heart of the argument by the Lind group and Hammes. The present work follows this line of reasoning with regard to this sort of conflict.

Another of Curtis's critiques is with respect to Hammes's inclusion of politics in the analysis of FGW that is inconsistent with theory and practice, meaning that tools like propaganda, the media, cyber warfare, and politics are emphasized. Curtis believes the actual combat of FGW should be examined. A second critique in this line of reasoning is that Hammes looks almost exclusively at guerrilla warfare (Curtis 2005). While this is true to some extent, Hammes sees FGW as more than just an evolution of warfare, since he includes the idea of signaling intentions which usually can be done through support of guerrilla movements and tends to be a state-to-state undertaking. Moreover, the emphasis on military struggle clearly puts Hammes in the FGW camp and is in line with the current situation between the US and Iran.

Echevarria Critique of FGW

Antulio Echevarria (2003) has written that FGW is a myth that attempts to supplant and muddle traditional theories of warfare, clouding these theories with semantic jargon. Echevarria contends that FGW is neither based on good history nor good theory. Echevarria has several specific critiques of FGW. The first critique is the theory's emphasis on high-tech weaponry, which is disputed by the records of Iraq Wars in 1991 and 2003. The second critique is of future war not being bound to the nation-state in some manner but initiated and sustained by non-state actors. The fact that history is littered with examples of total warfare is counter-factual evidence. Third is the claim that FGW is broader than other types of previous warfare. Echevarria believes that FGW fails to take into account that all terrorist organizations and insurgent groups are somehow funded or linked to the traditional nation-states (Echevarria 2003, 2005). This critique is correct to some extent but tends to ignore the Tamil Tigers or Al Qaida which are not controlled or funded by a specific nation-state. In fact, these groups have been successful in fighting against nation-states for many years. Echevarria's assertion that FGW does not account for Hamas or Hezbollah and their social networking and services is not totally correct.

Providing social services is not unique. The insurgency model of guerillas within society receiving assistance from the populous is not dated; it is merely not as prevalent as before since insurgent groups are becoming self-financing. The social roles taken by Hamas and Hezbollah are part of their overall ideology and part of the political aspects of FGW. In order to 'convert' the rest of the population to their cause these groups use politics and economics rather than just blending in among the population. This is a more effective strategy in the long run if control of the government is the war's aim.

Another critique of FGW theory is that it is a reasonable outcome of globalization or the globalization of conflict (Echevarria 2003). This is a misreading of FGW and avoids the inherent and interrelated political and economic aspects of this new type of warfare. FGW in its wider aspects is total or unrestricted warfare utilizing as primary tools—along with military means—politics and economics. To assert that these means are part and parcel of globalization is to overestimate globalization as a catalyst for conflict considering the major components of globalization are trade and financial movements. Echevarria notes that FGW theory seems to point to a 'super-insurgency' as the new norm yet this does not conform to historical record (Echevarria 2003, 2005). FGW is different and thus far we have not seen a 'super-insurgency.' However, the current wave of Al Qaida inspired terrorism and insurgency does suggest that ideologies can create movements and these movements can take on the characteristics of insurgencies both in ideology and actions. In sum, Echevarria criticizes FGW on the grounds that it is not state centered, overplays the role of insurgencies, and misses the point about historical examples. While this critique does point out some inconsistencies it does not shatter the basis of what we see in the long run low-intensity war between the USA and the Islamic Republic of Iran.

A final critique, implicit in the above, is the notion that any military innovation (be it technological, strategic, or tactical) will be subject to counter-innovation which can have the effect of eroding initial advantage eventually nullifying the innovation's advantage entirely. While this is certainly true in many instances, the evolutionary nature of fourth generation warfare makes it much more difficult to nullify any advantage one combatant may have. For example, Moore's Law notes how the computing power of microprocessors will grow exponentially. In the same manner, damage brought about by cyber-warfare will most likely grow. Both processes are evolutionary utilizing new tactics, technologies, and strategies. What makes fourth generation warfare and cyber-warfare similar is the fact that both can be done at varying levels of intensity and can be done from dispersed sources. The end is not the destruction of the opponent but the sapping of the opponent's will to continue the conflict through various means. It is this flexibility that makes fourth generation, low-intensity conflict unique and difficult to defend against.

Our Understanding of FGW and Low-Intensity Conflict

In this work we utilize the following understanding of FGW and low-intensity conflict. First, we find that there is indeed a new generation of warfare emerging yet we do not fully agree that it will be based upon principles of counter insurgency nor do we believe that it will be totally military based.

Second, we find that war aims are in line with the idea of a "moral victory". Through various avenues of coercion an enemy will be forced to change its mind or more precisely change its will to resist the opponent's wishes. Is this the same as previous iterations of coercion? In some ways it is different. Our definition of coercion (and the inherent aim of FGW) is to "subvert and change the will of a foe". FGW is encompassed by traditional theories of coercion yet furthered by the addition of several authors mentioned above who have detailed how modern FGW or low intensity conflict is fought.

Third, we do not exclude the modern concept and potentiality of the air-land battle, nor do we believe that insurgency warfare will always be the primary type of warfare in the future. Conversely we believe that a mix of both types—the conventional Air-Land battle of the Cold War mixed with insurgency— of combat will be viable in the future. The newly emerging multi-polar world ensures that maneuver warfare is alive and well while new nations and various international terrorist and irredentist groups ensure the continued use of insurgency and terrorist tactics to achieve political aims.

Fourth, we find that FGW is inherently a three-factor model utilizing the latest in technology and tactics to achieve the goal of demoralization and "changing the will" of the adversary. To this end the combatant is not just the military but policy, both diplomatic and economic, and (even in extreme cases) direct economic intervention by central governments to support war aims, such as economic embargoes or attempts to undermine a hostile currency or capital markets.

Fifth, we find evidence to indicate that the relationship between the Islamic Republic of Iran and the United States (since 1979) has been a FGW fought in a low intensity manner. This conflict has consumed significant human, economic, and political resources, yet has remained at a stalemate. The economic, political, and military capital that has been expended or lost in this conflict (either directly or indirectly in the form of forfeited economic gain) is but a harbinger of the types of war we will see in the future. With modern democracies not willing to accept massive casualties, warfare will become less intense and will move to new arenas (such as the economic or political arenas) to keep causalities low unless extraordinary conditions emerges. This strategy will also necessarily lengthen FGW low-intensity conflicts, yet will make them more politically palatable. Sometimes a state of low-level warfare is sufficient to keep the conflict alive as it can be used to cement domestic support for additional resources. In sum, we envision a fluid, low-intensity battlefield that encompasses political, military, and economic spheres.

The low-level conflict between the Islamic Republic of Iran and the United States began in 1979 and has continued to the present in roughly the same form. The ebbs and flows of the conflict have revolved around domestic politics in each nation, global politics, economics, and military means. At any given point each adversary chooses from political, economic, or military means to continue the conflict. In some instances, the reaction to one type of attack or "hostile initiative" may be in a different sphere of action than that of the attack. For instance, Iranian arming of Iraqi insurgents may not be met with military force but with diplomatic or economic pressure. Clearly, a different array of resources and tactics are available to decision makers and military leaders in this sort of low intensity conflict. The question remains: who decides what sort of strategy and tactics will be used and will these be in the economic, military, or political realms of action? To determine this one must look inside the political systems of the Islamic Republic of Iran and the United States.

Low-Intensity War and Domestic Politics

Military and political leaders ignore the domestic political situation of their adversaries at their own peril. While domestic political coalitions may be fluid they do indeed matter for politicians and those that politicians control (especially security and military policy decisions). Moreover, in a situation like Iran, the paramilitary forces (the Pasdaran) are rival to the regular military and a major economic and political player as well (Byman et al. 2001, Cordesman 2007, Roberts 1996, Schahgaldian 1987, Wehrey 2009).

To unravel how low-intensity warfare is carried out in the three spheres of action we look at the notion of the political selectorate or ruling coalition and coalition dynamics. The theoretical import for this study revolves around the fact that the conflict between Iran and the United States had been in motion for almost three decades and that the ebb and flow of the conflict can be directly tied to domestic politics in each nation. To this end we center the domestic part of fourth generation warfare analysis in this work squarely in the realm of domestic politics. The domestic political arena determines what tactics and strategies are used in each of the spheres of action in fourth generation, low intensity conflict.

The Selectorate and Low-Intensity Conflict

This section outlines how the selectorate influences the various decisions that elites make in pursuing conflict and low-intensity conflict in particular. All political leaders seek to maximize their tenure in office. Some resort to extra constitutional means (such as Alberto Fujimori did in Peru) while others lead via a one party authoritarian state as in Egypt. Domestic politics and domestic electoral groups ultimately determine the tenure of leaders. Selectorate theory by Bueno de Mesquita, Smith, Siverson, and Morrow (2003, 2008) assert that all nations have essentially the same institutional structure, in so far as they select and retain their leaders (Bueno de Mesquita 2008, 2010, Bueno de Mesquita et al 1999, 2001, 2002, 2003, 2004, 2005, Bueno de Mesquita & Ray 2004, Bueno de Mesquita & Siverson 1997, Bueno de Mesquita & Smith 2009).

Selectorate Theory Involves Various Elements

The Selectorate is a set of people in a nation-state who choose the representatives of the citizenry to form a government. In democracies the electorate chooses representatives through different open elections. Various demographic attributes of voters such as income, education, region, parent's political views, *etc.* can influence voting behavior. For example, in the United States the level of education tends to be correlated with voting for certain political parties. Thus, the selectorate at any given time will be those who prefer a particular candidate to another, based upon various factors. The selectorate is not fixed but fluid and can change its candidate preferences rapidly depending on the situation. The winning coalition is determined by the number of selectors a leader must have to remain in power. The winning coalition is drawn from the selectorate, or more specifically, from the various subsets of the electorate that support one candidate's policy preferences. The sizes of these sets are denoted as S and W selectorate for the former and winning coalition for the latter. The support coalition is the number of selectors who support the present leader and without their support the leader would fall. This is W or the winning coalition The support coalition is an aggregate of the subsets, or voting groups, of the selectorate that prefer the policy preferences of one candidate *vis a vis* another.

The leader needs the support of this coalition, and as such needs to be able to support those policies that various members of the coalition see as payoff for their support. A clever leader will ameliorate the demands of various coalition members either through compromise or by playing one versus another. To keep a balance the leader needs to distribute goods to the coalition members. These payoffs need not be material goods but can be policy preferences that assist the coalition members in some way. Examples are tax breaks or state contracts. If the size of the support coalition falls below the winning coalition level (or W) then the leader is in danger of removal or replacement by coalition members. Clearly the delicate balance leaders must maintain is of extreme importance as their position is dependent upon the continued loyalty and support of various coalition members. A challenger to the leader who is in peril must create a winning coalition of their own at least equal the size of the previous leader's winning coalition. The dynamics of coalition politics is such that, as mentioned above, a delicate balance must be maintained. In sum, to maintain power a leader must distribute goods to the various subsets of his winning coalition, failure to do this may render the leader vulnerable. Potential rivals must have at least the same size winning coalition to topple the current leader.

Democracies and Autocracies

The type of government determines the size of the selectorate, the coalition size, and thus the distribution of payoffs. Democracies have large selectorates (individuals and groups who support a particular leader) and large coalitions; thus, the tactics for potential leaders will be different and distribution of payoffs (and the payoffs themselves) will be different. Leaders in democracies, especially large democracies, have a finite amount of goods to distribute; thus, the size of the goods will quantitatively be smaller for each individual than in other systems.

Autocracies (one party states), on the other hand, have much smaller wining coalitions even though their selectorates may be as large as democracies. The strategies of the autocratic leader will be different in keeping the coalition together. The distribution of payoffs will be different from that of nations with large winning coalitions. Coalitions receive larger and different kinds of payoffs—government contracts, special export or import licenses *etc*. Monarchies and military dictators have small selectorates and wining coalitions, yet the same problem in distribution of goods plagues these governments as well. They have to distribute goods, albeit in a different manner.

Contrary to common belief, distribution of payoffs under dictatorships could be very difficult. For example, to retain the loyalty of the various Bedouin tribes in Saudi Arabia Ibn-Saud not only married numerous times to seal alliances he also made sure that funds were distributed to the tribes to maintain their loyalty. Military dictators may have to support the military in various ways including new weapons and promotions to ensure the loyalty of the forces. Each governmental form has a specific type of winning coalition and a corresponding logic to its distribution of payoffs. Failing to distribute goods according to the needs and wants of the specific coalition can end a leader's tenure in office or life.

Leadership and Distribution of Goods

In the distribution of payoffs the logic is very simple. Leaders remain leaders by satisfying their winning coalition. This is independent of the type of governmental system. Leaders can produce either public goods or private benefits via state policies. Public goods include civil liberties, sound economic policies, and national security. These are not excludable goods. Thus, in a typical post-election situation, leaders attempt to shore up their position by ensuring that economic expectations of the winning coalition are fulfilled and that foreign policy matters remain free of conflict or potential conflict.

While these general measures are important there may be more specific measures that appeal to the winning coalition. For example, some constituencies (such as stockholders) may receive capital gains tax relief. Such relatively painless measures can affect many and garner future support, especially from older voters who live off investment income and widely participate in elections (at least in the United States). This would be considered a non-excludable good since anyone who was a stockholder would benefit and the reduced tax treatment would apply to all equity holders. Conversely, private benefits are benefits or goods than can be excluded or apportioned to select persons or institutions. These involve economic gain, prestige, and special trading rights. While seemingly similar to the example above, private gains are excludable by virtue of their nature. For example, an excludable good would be an import license that is issued to an individual by an authoritarian leader. Holders of licenses stand to increase their wealth considerably. However, the license holder holds the license at the whim of the dictator. Usually some sort of kickback or percentage is given to the leader who has the power to revoke the licenses. Most states, democracies, authoritarians, monarchies, and dictatorships tend to produce both types of goods. However, as Bueno de Mesquita notes, "it is the mix of the two that varies with institutions (Morrow 2008)." This does have repercussions for the distribution of payoffs. As the size of the winning coalition increases the leader moves away from private goods to public goods since a larger coalition means more individuals and groups need to be appeased. This movement away from private to pubic goods affects the policy set leaders have at their disposal to remain in leadership roles. For example, an authoritarian or military junta may use export licenses to gain members of their coalition, while in a democracy coalition members may be attracted by preferential taxes on investments.

In a switch from an authoritarian system to a democratic system the private good of licenses that allows a few to gain considerable wealth may give way to tax changes. Tax changes allow larger numbers of people to be taxed less thereby solidifying not only a new base but also to hold on to the old licensees' if they are accorded the same preferential tax treatment. The flexibility enjoyed by leaders can seriously affect their tenure in office. The more flexible the incentive structure of goods, the more likely leaders will retain their offices. The ability to exclude or grant goods to various coalition members is relatively easy when the goods are intangibles like civil rights or are small like minor tax breaks. In the realm of forging policy an additional calculation applies within the overall framework of the selectorate.

The Selectorate and War

The selectorate has implications for countries engaged in fourth generation, low-level conflict. First, a large winning coalition forces the leadership to produce public goods. One of these goods is foreign policy success. Even a minor foreign policy victory can increase the popularity of a leader considerably, while also helping weak leaders shed their image of being weak. Second, small winning coalitions force a different calculus on a leader. In this instance failure in foreign policy will require private benefits to be distributed to the winning coalition to prevent defections. Third, in war (or a low-level conflict for that matter) leaders in a large winning coalition try to attain public goods. These may be victory, numerous small victories, or a perceived string of foreign policy successes aimed at the adversary. To keep the support of the selectorate the leader must ensure this public good is spread across the coalition to maintain support.

Fourth, in a small selectorate with the state engaged in a low level conflict, the leader must attempt to distribute private benefits to coalition partners to secure their loyalty in a low-level conflict. Moreover, it is important for the leader to be able to manipulate the flow of private goods to reward those parts of the coalition who favor, or are more supportive of, the conflict and be able to punish (or withhold) private goods from those parts of the coalition who do not support low-level engagement.

In sum, these four points demonstrate that a country pursuing low-level conflict within the fourth generation war framework will do so based upon the selectorate that chooses the nation's decision makers. Clearly, for both the US and Iran, the conflict is costly in economic terms but it can also be used politically to gain votes. From the perspectives of political elites in each country this is a rational strategy and both have used the conflict for their political advantage. Indeed politicians like to have an enemy that allows them to rally voters. Because it is an advantageous topic for politicians to utilize, as well as a relatively low risk issue (there is little chance of direct military engagement at this point in time), politicians in both countries use the issue of conflict consistently and liberally.

Implications for Iran and the US

The United States and Iran, as democratic nations, tend to gravitate to the extremes of the coalition scale. The United States has traditionally been a nation with a rather large winning coalition composed of many different groups whose power is diffused by the sheer size of the coalition. While interest groups can and do lobby the federal government the diffusion of power in the winning coalition coupled with the structure of the American constitutional systems ensure that a winning coalition is diverse and necessarily large. Moreover, this system has multiple entry points for those seeking rents or governmental favors thereby allowing more access. Yet, additional coalition-building is required to enact a policy preference. Payoffs in this arrangement are necessarily public goods. These public goods may be monetary as in tax cuts or increased services. However, in line with selectorate theory, these benefits are relatively small given the number of groups and individuals that receive at least part of the benefit. Given the smallness of the benefits these benefits tend to be in the foreign policy realm as predicted by selectorate theory. A US president can expect to gain about five percentage points in polls when some sort of foreign policy gain takes place, particularly a military strike (Bueno de Mesquita et al. 1999, Bueno de Mesquita & Ray 2004, Bueno de Mesquita & Siverson 1997). These types of foreign policy events reinforce the perception that the president is strong and engaged in protecting US interests. In the Iranian context there is less emphasis on the actual benefits that accrue from the continuance of conflict with the United States. The conflict is seen by most groups of the selectorate as a nationalist issue with deep roots in the humiliations Iran has suffered over the years. When Iranian politicians use the ongoing conflict with the United States it evokes nationalism in many respects rather than wholehearted support for the government. This is an example of how politicians can manipulate the selectorate by providing foreign policy successes or moral victories to hold the support of the selectorate.

A Modification of Selectorate Theory

Selectorate theory provides a parsimonious way of examining the decision making process in various forms of governmental systems. However, while it has good explanatory value in some respects, it fails to address intangibles that can frequently explain behaviors that seem less than optimal. The extended low-level of conflict between the United States and Iran is an example. This conflict cannot simply be explained by the selectorate, alone. However, its continuation for almost three decades can in part be explained by the selectorate. We believe that, in addition to selectorate theory described earlier, an intervening variable relating to the selectorate must be included. This variable is institutional or historical memory. In cases where the selectorate explanation fails to generate significant explanatory power the addition of the institutional or historical memory variable completes the model and makes the explanation more credible.

We integrate the idea of institutional memory in the following manner. According to selectorate theory a large winning coalition forces the leader to produce public goods such as victory in war or foreign policy victories. However, these public goods do not take into account the institutional and social history of a nation. In some nations the loss of territory or prestige is a non-tangible public good that cannot be measured immediately but can, in fact, impact decisions in the future. For example, the 'Vietnam Syndrome' in US foreign policy is an unstated policy of restraint in committing military forces based on previous experiences in the Vietnam War. Therefore, when making an analysis of US foreign policy using the selectorate as an explanatory theory, one must temper the predictions of the theory with that of the institutional or social memory. For example, when the Shah of Iran relinquished the island of Bahrain to Bahraini sovereignty, he provided private goods for his small selectorate yet he also created a negative public good (a public bad or negative externality) in that the vast majority of the Iranian population viewed Bahrain as an integral part of Iran. The Shah was seen as wrong in his relinquishment of Iranian claims to the island. The social or institutional memory of this event did not manifest itself until later when the Islamic Revolution was underway.

Clearly, institutional memory (historical as embodied in institutions) or social memory plays a large part in the way the selectorate views leadership and how such memory may trump conventional predictions of selectorate theory. For our purposes this addition to selectorate theory in decision making is important since conventional rational expectations of both Iranian and US selectorates cannot fully explain the reasons why both nations have engaged in a fourth generation, low-intensity conflict since 1979. Therefore, throughout this work, we temper the predictions of selectorate theory as applied to fourth generation, low-intensity warfare with an assessment of the institutional or social memory of previous actions. Such an analysis yields a richer, more rigorous explanation of this previously cooperative, yet now adversarial, relationship.

Implications for This Study

Given the nature of fourth generation warfare, and in particular the low-intensity version, decision making by leaders plays a large part in the success and the initiation of low-intensity actions and reactions. The balance of this work examines the basis and actions of low-intensity, fourth generation conflict being played out in the Persian Gulf by Iran and the United States. Specifically, we examine the major players, the winning coalitions over time, and how these changing coalitions have influenced US-Iranian relations and the low intensity conflict that both have been engaged in since 1979. The change over the years in the ruling coalition and how this has affected policies is examined in the various chapters of this work. The role of history, social movements, and the general level of global hostilities also play a part in the conflict between the United States and Iran.

The balance of this work focuses on the relationship between Iran and the United States in various facets. Chapter 2 examines Iranian-American relations from the early days to the end of the Pahlavi Dynasty. Written using many Iranian sources this chapter shows how colonialism and great power rivalry affected the early relationship between the United States and Iran. The cumulative effects of the push and pull of these early contacts was a distrust of American motives, even when they were benign. In many cases the best intentions put Americans at odds with Iranians which soured relations in several instances. Chapter 3 looks at the development of Iranian-American relations following the fall of the Pahlavi Dynasty to the present emphasizing how direct American actions on one hand furthered short term American interests while undermining long-term, post-revolution relations.

Chapter 4 takes an in-depth look at the economic relations between Iran and the United States and how this is also a forum of low-intensity conflict. Political and social factors are examined to show how economics and politics have driven the economic relations between the two countries. Chapter 5 looks at direct military confrontation between the two states and how that has driven politics and military policy in the past thirty years. Chapter 6 looks at specific incidences of Iranian and American military confrontations that are of a low-intensity. The struggle takes place in the Persian Gulf, in Lebanon and in subtle ways in other countries. Chapter 7 concludes this work by looking at ways the United States can deal with Iran especially an Iran with nuclear weapons capabilities, how that affects US politics, as well as how this would affect Iranian politics. We hope that this examination of Iranian-US relations through the lens of a low-intensity conflict will illuminate factors that are relevant for ending the hostilities and restoring relations on an equitable basis for the mutual benefit of both societies.

Bibliography

Abrahamian, E., 2008. *A History of Modern Iran*, NYC: Cambridge University Press.

Arquilla, J., Ronfeldt, D. & Zanini, M., 1999. Networks, Netwar, and Information-Age Terrorism. In Z. Khalilzad, J. P. White, & A. W. Marshsall *Strategic Appraisal: The Changing Role of Information in Warfare*. Santa Monica, CA: RAND Corporation, pp. 75-111.

Arquilla, J. & Ronfeldt, D., 2000. *Swarming and the Future of Conflict*, Santa Monica, CA: Rand Corp.

Barnett, R.W., 2005. Technology and Naval Blockade: Past Impact and Future Prospects. *Naval War College Review*, 58(3).

Bueno De Mesquita, B. & Siverson, R.M., 1997. Nasty or Nice?: Political Systems, Endogenous Norms, and the Treatment of Adversaries. *The Journal of Conflict Resolution*, 41(1), 175-199.

Bueno De Mesquita, B. et al., 1999. An Institutional Explanation of the Democratic Peace. *The American Political Science Review*, 93(4), 791-807.

Bueno De Mesquita, B. et al., 1999. Policy Failure and Political Survival: The Contribution of Political Institutions. *The Journal of Conflict Resolution*, 43(2), 147-161.

Bueno De Mesquita, B. et al., 2000. *Institutions, Outcomes and the Survival of Leaders*,

Bueno De Mesquita, B. et al., 2001. Political Competition and Economic Growth. *Journal Of Democracy*, 12(1), 58-72.

Bueno De Mesquita, B. et al., 2002. Political Institutions, Policy Choice and the Survival of Leaders. *British Journal of Political Science*, 32(4), 559-590. Available at: http://www.journals.cambridge.org/abstract_S0007123402000236.

Bueno De Mesquita, B. et al., 2003. *The Logic of Political Survival*, Cambridge: The MIT Press.

Bueno De Mesquita, B. et al., 2004. Testing Novel Implications from the Selectorate Theory of War. *World Politics*, 56(April), 363-88.

Bueno De Mesquita, B. & Ray, J.L., 2004. The National Interest Versus Individual Political Ambition: Democracy, Autocracy, and the Reciprocation of Force and Violence in Militarized Interstate Disputes. In P. Diehl *The Scourge of War: New Extensions on an old Problem*. Ann Arbor: University of Michigan Press, pp. 94-119.

Bueno De Mesquita, B. et al., 2005. Thinking Inside the Box: A Closer Look at Democracy and Human Rights. *International Studies Quarterly*, 49(3), 439-458. Available at: http://www.blackwell-synergy.com/doi/abs/10.1111/j.1468-2478.2005.00372.x.

Bueno De Mesquita, B. & Smith, A., 2009. Political Survival and Endogenous Institutional Change. *Comparative Political Studies*, 42(2), 167-197. Available at: http://cps.sagepub.com/cgi/doi/10.1177/0010414008323330.

Bueno De Mesquita, B., 2010. Foreign Policy Analysis and Rational Choice Models. In R. A. Denemark *The International Studies Encyclopedia*. International Studies Association.

Byman, D. et al., 2001. *Iran's Security Policy in the Post-Revolutionary Era*. Santa Monica, CA.

Cordesman, A.H. & Al-Rodhan, K.R., 2006. The Gulf Military Forces in an Era of Asymmetric War: Iran. *International Studies*, 1(202).

Cordesman, A.H., 2007. *Iran's Revolutionary Guards, the Al Quds Force, and Other Intelligence and Paramilitary Forces*, Washington, D.C.

Cordesman, A.H. & Seitz, A.C., 2009. Gulf Threats, Risks, and Vulnerabilties: Terrorism and Asymmetric Warfare. *Terrorism*.

Curtis, V.J., 2005. The Theory of Fourth Generation Warfare. *Canadian Army Journal*, 8(4), 17-32.

Dorman, A.M., 2008. *Transforming to Effects-Based Operations: Lessons from the United Kingdom Experience*, Carlisle, PA.

Dunlap, C.J., 2006. Neo-Strategicon: The Modernized Principles of War for the 21st Century. *Military Review*, (April), 42-48.

Dupuy, Trevor. Attrition: Forecasting Battle Casualties And Equipment Losses In Modern War, Virginia, 1990.

Friedman, Thomas (2000), *The Lexus and the Olive Tree: Understanding Globalization* (New York: Anchor Books).

Echevarria, A.J., 2003. *Globalization and the Nature of War*, Carlisle, PA.

Echevarria, A.J., 2005. *Fourth-Generation War and Other Myths*, Carlisle, PA.

Edwards, S.J., 2000. Toward A Swarming Doctrine? In *Swarming on the Battlefield: Past, Present, and Future*. Santa Monica, CA: Rand, pp. 65-85.

Edwards, S.J., 2005. *Swarming and the Future of War*, Santa Monica, CA: Rand.

Fairbank, J.K., 1986. *The Great Chinese Revolution, 1800-1985* 1 ed., New York: Harper & Row.

Flanagan, S. & Schear, J., 2008. *Strategic challenges: America's global security agenda*, Washington, D.C.: National Defense University Press. Available at: http://www.potomacbooksinc.com/resrcs/press/1597971219_pressrel.pdf.

Ganguly, S., 1996. Explaining the Kashmir Insurgency: Political Mobilization and Institutional Decay. *International Security*, 21(2), 76-107.

Ganji, B., 2006. *Iran & Israel: Asymmetric Warfare and Regional Strategy*,

Gordon, S.L., 1998. *The Vulture and The Snake Counter-Guerrilla Air Warfare : The War in Southern Lebanon*, Begin-Sadat Center for Strategic Studies.

Hajjar, S.G., 2002. *Hizballah: Terrorism, National Liberation, or Menace?*, Carlisle, PA: Strategic Studies Institute.

Hammes, T.X., 2005. Insurgency : Modern Warfare Evolves into a Fourth Generation. *Strategic Forum*, (214), 1-8.

Hammes, T.X., 2007. Fourth Generation Warfare Evolves, Fifth Emerges. *Military Review*, (May-June), 14-23.

Hoffman, F.G., 2009. Hybrid Threats : Reconceptualizing the Evolving Character of Modern Conflict. *Strategic Forum*, (240), 1-8.

Hugill, P.D., 1998. The Continuing Utility of Naval Blockades in the Twenty-First Century.

Jordet, N., Explaining the Long-term Hostility between the United States and Iran : A Historical, Theoretical and Methodological Framework. Middle East.

Kinross, S., 2004. Clausewitz and Low-Intensity Conflict. *Journal of Strategic Studies*, 27(1), 35-58.

Krepinevich, A., Watts, B. & Work, R., 2003. Meeting the Anti-Access and Area-Denial Challenge. *CSBA Report*.

Larson, E.V. et al., 2004. *Assuring Access in Key Strategic Regions: Toward a Long-Term Strategy*, Arlington, VA: RAND Corporation.

Liang, Q. & Xiangsui, W., 1999. *Unrestricted Warfare*, Beijing: CIA.

Lind, J.M., 2004. Pascifism or Passing the Buck: Testing Theories of Japanese Security Policy. *International Security*, 29(1), 92-121.

Lind, William S., Nightengale, Keith (Colonel, USA), Schmitt, John F. (Captain, USMC), Sutton, Joseph W. Colonel, USA), Wilson, Gary I. (Lieutenant Colonel, USMC-R). (October 1989), 'The Changing Face of War: Into the Fourth Generation', *Marine Corps Gazette,* 22-6.

Lind, W.S., 2004. *Fourth Generation War*, Imperial and Royal Austro-Hungarian Marine Corps.

Lock-Pullan, R., 2005. How to Rethink War: Conceptual Innovaiton and Airland Battle Doctrine. *The Journal of Strategic Studies*, 28(4), 679-702.

Mao, T., 2000. *On Guerilla Warfare* 2nd., Champaign, IL: University of Illinois Press.

Metz, S., 2000. *Armed Conflict in the 21st Century: The Information Revolution and Post-Modern Warfare*, Carlisle, PA.

Moore, Harold G.; and Galloway, Joseph L. (1992) *We were Soldiers Once..And Young: Ia Drang--The Battle That Changed The War In Vietnam* (1ˢᵗ ed.). Random House. ISBN 0-679-41158-5.

Morrow, J.D. et al., 2008. Retesting Selectorate Theory: Separating the Effects of W from Other Elements of Democracy. *American Political Science Review*, 102(03), 393-400. Available at: http://www.journals.cambridge.org/abstract_S0003055408080295.

North, R.C. & Choucri, N., 1983. Economic and Political Factors in International Conflict and Integration. *International Studies Quarterly*, 27(4), 443-461.

Pollack, K.M., 2004. The Persian Puzzle: The Conflict Between Iran and America, New York: Random House.

Putnam, R.D., 1988. Diplomacy and Domestic Politics: The Logic of Two-Level Games. *International Organization*, 42(3), 427-460.

Quosh, C., 2007. *American Foreign Policy Towards Iran: Between Values and Interests or Beyond?*, Hamburg.

Roberts, M., 1996. *Khomeini's Incorporation of the Iranian Military*, Washington, D.C.

Schahgaldian, N.B. & Barkhordarian, G., 1987. *The Iranian Military Under the Islamic Republic*, Santa Monica, CA: Rand.

Schahgaldian, N.B., 1989. *The Clerical Establishment in Iran*, Santa Monica, CA: Rand.

Schahgaldian, N.B., 1994. *Iran and the Postwar Security in the Persian Gulf*, Santa Monica, CA.

Singer, P.W., 2002. Corporate Warriors: The Rise of the Privatized Military Industry and Its Ramifications for International Security. *International Security*, 26(3), 186-220.

Smith, L., 2007. *Iran, Hizbullah, Hamas, and the Global Jihad: A New conflict Paradigm for the West* L. Smith, Jerusalem: Jerusalem Center for Public Affairs.

Tangredi, S.J., 2000. *All Possible Wars ? Toward a Consensus View of the Future Security Environment, 2001-2025*, Washington, D.C.

Thaler, D.E. et al., 2010. *Mullahs, Guards, and Bonyads: An Exploration of Iranian Leadership Dynamics*, Santa Monica, CA: Rand.

Truver, S.C., 2008. Mines and Underwater IEDS in U.S. Ports and Waterways: Context, Threats, Challenges, and Solutions. *Naval War College Review*, 61(1), 106-127.

Wehrey, F. et al., 2009. *The Rise of The Pasdaran: Assessing the Domestic Roles of Iran's Revolutionary Guards Corps*, Santa Monica, CA: RAND.

Chapter 2

Early Contacts

Introduction

This chapter provides a brief history of US-Iran relations from the earliest days to the end of Qajar Dynasty. The main focus is historical. The relations between nations are complicated and multifaceted. Each component provides a different characteristic of the relations. The combined effect of the pieces shapes the actual relations between the two countries. Sometimes the parts function independently and at other times they are orchestrated. Nevertheless, the parts always interact and influence each other.

The beginning of the relationship between Iran and the United States was complex and dates back to a long time ago. It was shaped by those who were in power and those who brought them to power. Furthermore, the historic setting also plays an important role as history is played out through institutional and social memories that shaped decision making in the past and shapes decision making today.

An underlying social memory seen throughout Iranian history is its sense of empire lost and its identity as a crossroads of civilizations that have laid claim to much of its territory. These two institutional or social memories played out in the political and social history of the eighteenth century when contact with the United States initially began. Since the Iranian state was just beginning to coalesce with the Qajar Dynasty's decline, political and social elite tended to be one and the same supporting their overall positions of privilege within the state apparatus. The closed political system excluded most citizens and the elite loyal supporters of the monarchy were given preferential treatments, both economic and social. The clergy remained ensconced in its studies only venturing out when policies ran contradictory to Islam. The Shi'a clergy at this time was firmly against intervention in governmental affairs and resisted attempts to bring them into the political dynamic. In all, the end of the dynastic period of Iran was much the same as in other nations with a declining monarchical system, power became concentrated, authoritarian élites controlled the bureaucracy, and the military and the masses were excluded.

End of Qajar Dynasty and a Brief Summary

The Qajar dynasty was founded by Agha Mohammad Khan in 1779 ending a rough and turbulent period in Iran's history. The Qajar Dynasty ended in 1925. The fall of the last Qajar ruler, Ahmad Shah, marks the end of a period of profound decline in Iran's history. During this era, many European countries (especially Britain, Russia, and the Ottoman Empire) rose to the peak of their respective powers. By virtue of geographic proximity, the Ottoman Empire and Russia played an important role in Iran's demise. By virtue of its colonial rule over India and Persian Gulf territories, Britain also played an important role in Iran's decline (Lorentz 2006).

Iran was the last major non-European power in the region. In the years preceding the Qajar Dynasty, Iran simultaneously expanded its borders, and displayed its appetite and ability for conquest by attacking and capturing surrounding areas. Thus, Iran provided a vast territory with many riches that could be taken by these three superpowers. While Western countries were indefatigable in terms of development and expansion, Iran was in a state of arrested development: Iranian kings preferred pleasure seeking and fighting with rivals. Additionally, development and expansion for Iran was stunted by public servants; they advanced their careers and secured a comfortable living by taking an opportunistic, and at times unethical, approach regarding their office.

During 146 years of Qajar rule, Iran lost substantial regions with numerous riches and considerable geopolitical value. Russia, the Ottoman Empire, and Britain gained land from Qajar-era Iran. Currently, territories that were once part of the Qajar Empire are within the borders of Armenia, Azerbaijan, Turkmenistan, Afghanistan, Bahrain, and Pakistan. The treaties of Gulistan (1813), Turkmenchay (1828), and the Anglo-Russian Entente (1907) are undoubtedly the most infamous treaties imposed on Iran in recent history. Because of the first two treaties, Iran lost substantial territories to Russia. The Anglo-Russian Entente effectively divided Iran into three regions. The north became Russia's region of influence, the south became Britain's region of influence, and the center became a "neutral" buffer zone between the two powerhouses. Concessions to the Ottoman Empire and other land losses also dotted this dynastic legacy (Lorentz 2006: 13-15, 168-171, Daniel 2001: 105-106).

During this era the US had no ties with Iran. The Monroe Doctrine shaped US policy in such a way as to emphasize US influence in the Americas over US influence elsewhere. In 1823, President Monroe declared that foreign intervention in the newly independent countries of the Western hemisphere (by any European country) would be considered, by the United States, as a hostile act (Hart 1914, United States 1903). This is a bold statement from a leader whose country was just 62 years old. However, it also demonstrated that the United States did not consider itself strong enough to lock horns with European powers in the rest of the world.

One of the most hated components of the Turkmenchay Treaty was the privilege of capitulation for Russian citizens in Iran. Later, Britain demanded in-kind reciprocal treatment for its own citizens. The British argument was based on prior agreements. Those UK-Iran agreements stipulated that any right given to other countries must be extended to Britain as well. Later, in the 1960s, the United States obtained the same privilege with great consequences.

The word "capitulation" is the Latin equivalent of the Arabic-Iranian combination word "Ahdnameh". The word "ahd" in Arabic means "promise, convent, or treaty." The word "nameh" in Farsi means "letter, deed, or certificate." Therefore, "ahdnameh" means a certificate of a promise. When exchanged between two or more countries, this is what is known as a "treaty". For the Ottoman Empire, capitulation-agreements with European countries were also referred to as "Ottoman Charters." Most of the "ahdnamehs" were signed between the Ottoman Empire and Venice after the 1453 capture of Constantinople by the former. However, there is one such treaty dating back to 1403 between the two nations.

The "Ahdnamehs" go back in history to the glory days of Islam. Caliph Harun al-Rashid "imposed" capitulation to the subjects of Charlemagne in the ninth century. In those days the idea was viewed differently. The purpose of this law was to exclude foreigners from Islamic laws, a practice that went back to the days of the Prophet Mohammad. If a country agreed to pay taxes to the Islamic government it was allowed to keep its religion (Christians and Jews) and its own laws, but its citizens were denied the "privilege" of being citizens of the Islamic country. At least one European country, the Netherlands, gave similar promise to citizens of another European country—Portugal—in 1641. However, during colonial times, this became a symbol of dominance for the colonizer. Such covenants were seen to give an open reign to colonial citizens, employees, and military personnel to do as they wished without regard for local law. In most cases, the offending colonist received much lighter sentences under the court system. The colonial citizens were subject to law that favored them and lacked regard for laws in the region.

In Iran, the capitulation right was viewed as a symbolic surrender of self-determination and an acquiescence regarding foreign rule. This is a very sensitive issue for Iranians. History has shown that, for Iran, capitulation always ended in great resentment and social upheaval. Iranian's view of capitulation seems to be shared by western countries, none of which has ever granted such rights to other countries. For example, all foreigners in the United States are subject to the laws of the United States. The only exception is diplomatic immunity, which follows a different set of long standing diplomatic norms. The US does not grant domestic power to foreign nations or their citizens.

Throughout the second half of the reign of Qajar, and especially towards the last few decades, the people of Iran were struggling to gain their independence and self-respect. The Tobacco protests of 1892 were major events that contributed to the end of the Qajar dynasty. These protests were rooted in public outrage over concessions that the Shah granted Britain in 1890. Ayatollah Mirza Hassan Shirazi issued a fatwa banning the use of tobacco in December 1891. Many people obeyed the fatwa and the protest spread. Finally, the concession was revoked and the unrest eased. However, the people had managed to unite and flex their muscles (Bakhash 2009, Daniel 2001, Lorentz 2006, Shuster 1912). The forces supporting constitutional elections gained strength. Finally, in October 7, 1906 the opposition revolted and defeated the supporters of monarchy and established the first parliamentary election in Asia. The Iranian Parliament, named Majlis, continued pressuring the Shah for more reforms. Fearing they would lose control of Iran, the British and Russian governments signed the Anglo-Russian Entente of 1907. This effectively divided what was left of the country. However, the constitutionalists continued their pressure and demanded financial reform for the government.

At this time the country was practically bankrupt due to revolution; the Shah's extravagant trips to Europe for vacations and medical treatment; and abuses of Russia and Britain. The country had been borrowing heavily and was in bad financial shape. French financial advisers were hired, but they proved to be ineffective. Iran turned to Japan next. Japan had gained notoriety and admiration after it defeated Russia in 1905. That 1905 victory curbed Russian expansion in the Pacific region (Greaves 1968). Furthermore, the Japanese victory provided a glimpse of hope for Iran and other such countries that were being taken apart, piece by piece, by Russia, Britain, and other colonial powers of the era. However, there was no formal contact between Iran and Japan. Few attempts to obtain formal agreements proved futile. Therefore, Iran turned to the United States.

The US was the only country to have defeated Colonial Britain, freeing itself in the process. Although there was ample evidence to the contrary (especially in the Philippines and Central and South America) the United States was appealing because of its repeated confirmation of non-involvement in other nations' affairs. The US was reluctant to help Iran because they would risk going against British or Russian interests. Both empires put pressure on the US, and it looked as though US-Iranian negotiations were on the verge of collapse. However, President Taft intervened by ordering a special envoy be sent to Tehran. The envoy included Morgan Shuster, and 16 other experts (Abrahamian 2008, Bonakdarian nd, Ghanea Bassiri 2002, Greaves 1968, United States 1911). Shuster had established his credentials by serving at the Cuban Customs House from 1899 to 1901. He became the collector of customs at Manila, the Philippines, which was a US colony from 1896 to 1946. In 1906, he was appointed Secretary of Public Instruction in the Philippines. Shuster had served very well in the Philippines. In fact, some believe that he served too well (Abrahamian 2008, Bonakdarian nd, Ghanea Bassiri 2002, Greaves 1968, United States 1911). He was very strict, which made many people unhappy and resulted in his removal. In May 1911, Shuster (and his advisers) arrived in Tehran and became the Treasurer Advisor of Persia. Shuster and his advisers were actually employees of Iran's government. The United States government was only an intermediary that helped the two parties meet. One of the first things that Shuster did was to investigate the accounting records at the customs in order to centralize the customs affairs. In 1900 and 1902 Russia had given loans to Iran, for which the revenues of the customs as collateral (except for the customs of the southern part of the country, which was used as collateral for loans from Britain). Shuster's investigation revealed inconsistencies in the bookkeeping, which deprived Iran from its lawful revenue. The Russians were unhappy with Shuster's investigations and when Shuster confiscated the properties of the Shah's brother, who was working for Russia; Russian troops landed at the Port of Anzaly and demanded an official apology from the Iranian government (Mahdavy 1387

(2008), PP. 339-341). Under pressure from Russia and Britain, Shuster was ousted in December 1911, and left Iran by early 1912— after only few months at the job (Abrahamian 2008, Bonakdarian nd, Ghanea Bassiri 2002, Greaves 1968, United States 1911). Later that year, the Russians shelled the Goharshad Mosque in Mashhad (Abrahamian 2008, Bonakdarian nd, Ghanea Bassiri 2002, Greaves 1968, United States 1911). It is beneficial to note that although both Japan and the United States were Imperialists by the end of the nineteenth century, neither one had any colonies in the Middle East. Countries like Iran considered them a possible counterweight and ally in their struggle against the dominant colonial powers that operated in the region. As is evident from the above account, Iran, as a developing political and economic state, was dependent on her more powerful neighbors. In many instances larger powers attempted to exert influence over Iran and attempted to manipulate the Iranian government visa-a-vis the other powers.

Early US-Iran Relations

Different aspects of the relationship are addressed separately. The first section is devoted to political relations. The second section discusses missionaries and private citizens. This second section only addresses early private contact, when there was little or no official contact. Over time, and with the strengthening of official ties, the majority of these contacts between the two countries were channeled through official protocol. Furthermore, with the increase in relationships, the numbers of private citizens engaged in each other's countries became numerous, which peaked right before Iran's Islamic revolution. Therefore, except for the early days, private citizen conduct and engagement are not addressed unless they contribute substantially to the official US-Iran relationship either positively or negatively. Section three addresses the historical economic relations between the two countries. Section four addresses the early days of economic relations with a considerable discussion of oil. Section five briefly discusses cultural relations.

Vulnerability to colonial powers can be seen during this period through Iran's attempt to balance the power of one colonial power against another. This precarious game was not unnoticed by the vast majority of the population who, at times, had to deal with occupying foreigners. Ignorant of Iranian customs, foreign occupiers created a sense of hostility within the greater Iranian population. Powerful outsiders were known to be disdainful of local Iranians and cared little for the welfare of the nation as a whole. The first contacts between the United States and Iran were cordial but also reflect an underlying agenda on the part of the United States. To be sure, as a democracy, the US selectorate supported the old idea of "Manifest Destiny" whereby many felt that the United States destiny was to spread mainline Protestant Christianity, democracy, and capitalism to the far reaches of the globe. Supporters of such an agenda included many religious citizens who supported missionary activities as a way to spread their particular form of Christianity. Iran was seen as an Islamic nation in need of conversion despite the presence of the Armenian Apostolic Church, the Assyrian Church of the East as well as the Chaldean Catholic Church. Other supporters included the business class who sought commerce or access to raw materials. Foreign policy elite sought ways to counter Russian or British moves in the region as America sought colonies to compete with the earlier European colonial powers. In sum, the US introduction to Iran was one that included a definite social, economic, and political agenda.

Political Relations between United States and Iran

Early Days to 1953

During the eighteenth century, regional diplomacy was the focus of Iranian foreign policy while international diplomacy was primarily a concern of Colonial empires. Even during the nineteenth century, Iran's diplomatic ties were predominately with neighboring states. Iran was unique in that the Russian and Ottoman empires were adjacent.

Additionally, British rule over India made the British Empire a de facto neighbor. During this time, Iran had lost its not-so-distant glory and the Ottoman Empire was in decline. Decline for these countries coincided with the increase in Britain's influence in the southern territory of the Ottoman Empire and in Iran. Additionally, the rise of Russian influence and the eventual capture of vast area of Iran by Russia were noted earlier. Consequently, the majority of Iran's diplomatic efforts were directed towards Russia and Britain. Iran did not have much contact with other European powers.

During this period, Britain pursued several policies in the region. The first and the most important policy was to protect India. During Nader Shah's rule (1736-1747), Iran ransacked India. Also, Russia seemed to be taking an interest in expanding into India. Nader Shah, the founder of the Afsharid Dynasty, had invaded India several times beginning in 1738. Nader's successful attacks destroyed the Mogul Dynasty of India and may have paved the road for British colonialism (Daniel 2001).

The British East India Company was established in 1708. British merchants and military were motivated to do anything to weaken Iran and ensure that it could not attack India again. This policy eventually resulted in separation of Afghanistan from Iran in 1857 (Abrahamian 1969, 1979, 2008, Bonakdarian 1991, Daniel 2001, Lorentz 2006). Other parts of southeast Iran were separated from Iran's Baluchistan province for the same purpose. Britain also infiltrated the Persian Gulf; establishing a naval base in Bahrain (Daniel 2001). Eventually Britain controlled the southern part of Iran. By creating the Southern Police in Iran, Britain pursued its imperialistic ambitions and effectively limited the possible expansion of the Ottoman Empire into the southern parts of Iran. The Ottoman Empire was weakened by limiting its incursions to the northern parts of the country which created worries for the Russians only. Furthermore, Britain was creeping into southern regions of the Ottoman Empire by creating unrest in regions that today are known as Egypt, Syria, and Iraq. At the time, these countries, or parts of them, were all part of the Ottoman Empire. Some regions in today's Syria, Iraq, Armenia, Azerbaijan, and Georgia were continuously changing hands between Iran and the Ottoman Empire, and/or Russia depending on which nation (Iran or the Ottoman Empire) had the least inept ruler at the given time. During this era, the Russians were the main winners of almost all conflicts in that region.

The First Known Contacts between Iran and United States

On the one hand, Europe's constant conflict with the Ottoman Empire made them a natural ally of Iran from the thirteenth to the eighteenth centuries. On the other hand, their colonial aspirations created conflict over territories that either belonged to, or were of interest to, Iran. As a consequence of conflicting interests, official, modern diplomatic relations necessarily began between Iran and European countries dating back to the sixteenth century (Ferrier 1973, 1976, 1986).

The newly independent United States had limited global contact outside of Europe and Latin America. The establishment of the Monroe Doctrine in 1823 did not help expand diplomatic relationships with other countries either. However, this does not mean there was no contact between Iran and US. In fact, on October 19, 1851 a friendship and shipping agreement was signed between the US and Iran. The US representative to Iran was Gorge March. The Iranian envoy to the US was Mirza Ahmad Khan Khoyie. This treaty gave Americans a permit to open a consulate in Bushehr located in a strategic part of the Persian Gulf on Iranian shores. Until then, only Britain had such a privilege. However, this agreement was never carried out and no formal relationships were established.

Prior to the British invasion of Eastern Iran in 1839, the empire coerced the chiefs of Qandahar and Herat into signing treaties that ceded authority to the Crown. These treaties would eventually lead to the *de facto* and complete separation of the rest of Afghanistan from Iran, in 1940 (Brobst 1997, Daniel 2001, Metz 1989, Volodarsky 1985). In 1856, Iran put Herat under siege hoping to reclaim the disputed territory. Of course, by then, most of Afghanistan was firmly in the hands of local chieftains that were on Britain's payroll. The rest, mostly in Central Asia, were safely tucked under the Russian Bear by this time. On November 1, 1856, Britain declared war on Iran, and set about occupying a number of ports and cities in Persian Gulf.

Due to the imminent threat to the heartland and the capital, whatever troops Iran could muster were kept busy in Heart (Brobst 1997, Daniel 2001, Volodarsky 1985). The Iranians had not forgotten the Afghan warlords' attack of 1771 and the pursuant occupation that lasted until 1779. As the conflict depleted Iran's national wealth, Tehran approached Napoleon III of France in order to end the hostilities. Britain asked France to be the intermediary, for its own reasons. Britain and Iran signed a treaty in Paris on March 4, 1857. According to the treaty, Iran gave up all its rights to the "countries of Afghanistan" in return for cessation of hostilities and withdrawal of the British army from southern cities and Persian Gulf ports (Brobst 1997, Daniel 2001, Thornton 1954, 1955, Volodarsky 1983). After a humiliating defeat and the capture of its ports, Iran (through three of its consulates located in Istanbul, St. Petersburg, and Vienna) contacted the United States and requested assistance in safeguarding its southern ports and boarders. This is evidence that, despite the republic being so new, the United States was becoming a major international player whose assistance and influence were sought by weaker countries.

These events brought the United States and Iran closer together. On June 13, 1856 they signed an agreement. This initial agreement was signed to protect US citizens and facilitate Iran's access to either a naval fleet or naval protection from the US. Later that year, the two countries established embassies in each other's countries. The first US diplomat to serve in Iran was SGW Benjamin (1883-1885). He was first appointed as Minister of the American legation and later, he was appointed to the office of Minister Resident. Diplomatic relationships between the two countries were limited and unfruitful for many years. Benjamin was charmed by the "eastern allure" (Saleh, 1355 (1976), P218) of Iran and did not tend to anything that could be considered diplomatic in nature (Abrahamian 2008, Ghanea Bassiri 2002, US State Department 2001, Volodarsky 1983). His successor, Bayless Hanna, never made it to Iran. The next Minister Resident Hampden Winston, stayed for two months, and concluded that there was no trade benefit or potential. His conclusions were based on the lack of property rights; the lack of political leadership; and the rapid and continuous decline of Iranian currency.

The first US ambassador, Leland Morris, was not assigned until 1944. This was after the Tehran Conference which established Iran as a supply line to the Soviet Union and conferred upon Iran the epithet of "The Victory Bridge" (Abrahamian 2008, Davis 2006, Iran-USSR-Great Britain 1942, Ladjevardi 1983, US State Department 2001). Official contacts between the US and Iran lagged behind private contacts, as is customary. Iran's concern was due to dominance of Britain in the Persian Gulf, which was effectively demonstrated during the siege of Heart. Iran was also aware of the importance of a modern navy in light of the Crimean War of 1853-1856. It is worth mentioning that Iran's first diplomatic envoy to the United States was an ambassador named Mirza Abolhasan Shirazi, who arrived in Washington D.C. in 1856. This indicates how Iran valued its relationship with the new budding global power of the nineteenth century. Iran had high regards for the United States and hoped the United States would help Iran free itself from the domination of Russia and Britain, as several countries such as France, Holland, and Spain did for the United States in 1776.

However, the United States closely followed the Monroe Doctrine. The doctrine was first laid out during the Seventh State of the Union Address as given by President Monroe to the US Congress. The doctrine included non-participation with or against European powers, and the pursuit of (open) trade. After American diplomats decided that there was little potential for commerce with Iran, bilateral relations remained limited to the protection of United States citizens in Iran.

Diplomatic relationships between the two countries were sporadic for many years. Contributing factors were distance; slow transportation and communication; financial difficulties (Iran); and non-interventionist policy (United States). The efforts of Spencer Pratt (1866-1891) and Alexander McDonald (1893-1897) on behalf of the United States were the only two exceptions. These diplomats substantially improved the relationship, understanding of each other's countries, and diplomatic contacts. Nevertheless, the latter part of the nineteenth century witnessed limited diplomatic relationship between the two countries. The primary role of US envoys was to look after Presbyterian missionaries active in the Christian enclaves of northeast Iran in Azerbaijan (Abrahamian 2008, Mahdavi 2005, Seward 1912). During this period the extent of the support of Iran's embassy for Iranians in the United States is not clear.

The last years of the nineteenth century witnessed more internal turmoil in Iran. The country's decline; loss of territories; national humiliation caused by the imperial presence of Britain and Russia in the country; lack of law and order; and an overall sense of frustration with the ruling dynasty were all factors preparing the ground for a revolution. During these years, the relationships between the United States and Iran were improving, due in part to the efforts of Herbert Bowen and Mofakhamaldole (Minister Residents at their respective embassies).

The improved relationship could have been, in part, due to President McKinley's (the 25th president of the United States) imperialist views. These views were fueled by the defeat of the Spanish fleet in the 100-day war in Santiago Harbor, Cuba; the seizure of Manila, Philippines; and the invasion of Puerto Rico. President McKinley's murder in 1901 did not end the relationship between Iran and the United States, but actually improved it. A memorial service for President McKinley was set up in Tehran by the Iranian government. Most of the dignitaries participated. News of the memorial made a big impact in the United States.

The next US Minister Plenipotentiary Lloyd Griscom (1901-1902), spent most of his time traveling around Iran gathering information about economic resources and other strategic information. The tenure of his successor, Richmond Pearson (1902-1907), coincided with the beginning of Iran's Constitutional Revolution. In spite of substantial coverage of the events in US newspapers, the official US stance was non-interventionist. A more substantial event was that of the Anglo-Russian Convention of 1907. By this agreement, Iran was divided into three regions. The northern region was under Russian influence. The southern region was in Britain's control. Allegedly, the central region was left in the hands of Iranians (Abrahamian 2008, Bonakdarian nd, Lorentz 2006, The Recent Anglo-Russian Convention 1907).

In spite of an official non-interventionist approach of the United States government, missionaries and other US citizens supported the constitutionalist revolutionaries. To take care of its' citizens in the region, the United States opened a consulate in Tabriz, the heart of the Constitutional Revolution in 1906. The official non-interventionist approach of the United States lasted at least until the events related to Morgan Shuster.

To put things in perspective, the Panama Canal began in 1903 and was completed in 1914. The Panama Canal established the beginning of US participation in global imperialism. During the latter part of the nineteenth century, global imperialism was limited to Britain, France, Russia, and Germany. US participation in World War I wiped out any remaining doubts about the global power and increasing role in international affairs of the United States. From this point on, a major objective of the United States was to obtain cheap raw materials and secure markets for its finished products (Mojani 1384 HS 2005 AD). Before World War I (WWI), the United States owed $3.7 Billion to other countries. By 1925, Britain, France, and Soviet Union owed $21 Billion to the United States. In 1920 the US share of global exports was one sixth, and that of imports was one eighth. This brief highlight demonstrates how rapidly the United States became a superpower.

The turmoil of revolution; ineptness of the ruling class; British and Russian interventions in all aspects of Iran's internal affairs; and financial problems had bankrupted the country. The treasury was empty, and the king of Iran had to borrow money to tour Europe (Abrahamian 2008, Avery & Simmons 1974, Bostock 1989, Brockway 1941, Greaves, 1965a, 1965b, Wilson 2002). Therefore, in December of 1909, the Majlis (Iran's constitutional parliament) sent a request to the United States government. The request was for a qualified person to oversee Iran's finances. In May 1911, Morgan Shuster was appointed as treasurer general of Persia. Under pressure from Britain and Russia, he was forced out of Iran in less than a year.

When World War I began and the Germans achieved rapid gains against the three powerful colonial powers (Britain, France, and Russia), Iranians were excited. The public was demanding that Iran join the Central Powers and use that alliance to get rid of colonial superpowers that had been tormenting the country for over 150 years. In fact, Ottoman's Sheikh al Islam gave a Fatwa for Jihad against Britain, France, and Russia, which were the core of the Entente Powers, but the Iranian government and people ignored the Fatwa.

Citizens Abroad

United States Citizens in Iran

The first known American citizens in Iran were Harrison Gray, Otis Dwight, and Eli Smith. They arrived in Tabriz on December 18, 1830 as part of the Presbyterian Missionary. Within two years, another student by the name of Meriek from the same theological school arrived in Iran. He and two Germans traveled to Tehran, Isfahan, and Shiraz. Based on their recommendation, the American Board of Commissioners for Foreign Missions sent a Minister named Justin Perkins and a physician named Asahel Grant. They established the first permanent missionary center in Orumiyeh. The center included a library, hospital, school, and a print shop. According to Mojany (1384 HS (2005), P P34-36), the Americans were able to grow their mission in spite of insulting Islam and converting people to a new foreign religion Mojany (1384 HS (2005), P P34- 36). A contributing factor was that, at the time, US citizens in Iran were protected by Britain (Abrahamian 2008, Lorentz 2006, Mahdavi 2005, Zirinsky 1993). By 1873, another religious school was established in the southern part of the capital Tehran. These schools operated under the umbrella of spreading science but were actually proselytizing. Finally, due to widespread religious teaching and subsequent complaints, Naseraldin Shah forbade the establishment of any new school. The Shah's decree was ignored, and in 1881, another school was established in the northern part of Tehran. This school had the first artisan well, which was its "science" contribution (Archive of (Iranian) Foreign Ministry 1301 HG (1884 A), Notebook 118, P 23). Samuel Jordan established the first American college in 1925, some 27 years after his arrival to Iran (Abrahamian 2008, Ghanea Bassiri 2002, Lorentz 2006, Mahdavi 2005, Zirinsky 1993). Although Jordan was well liked by Iranians the activities of other "educators" behind the scenes were offensive to Iranians (Mojany, 1384 HS (2005), P P38). The objections were more concentrated in the northwestern parts of the country where the missionaries were seeking to convert local citizens more aggressively, which created resentments, and at times, hostility by the locals. These feeling and hostilities sometimes culminated by protesters entering into missionary buildings (Mojany 1384 HS (2005), P 40).

Another motive of these missionaries was to gain political influence for the United States, which would come through admitting the children of the elite into the missionary schools (Powell 1923, Mahdavi 2005, Zirinsky 1993). At the time family ties were crucial for obtaining and securing high ranking governmental posts. The attempt to influence the culture and the minds of the people was the use of foreign names and words. For example a physician named David W. Torrance established a hospital in Tehran on a 37 acre lot (15 hectares) named "West Minister." In 1904 a disease outbreak in Iran led to the creation of four mobile hospitals in different parts of Tehran by the missionaries. The missionaries were using extended mobility to identify areas and people that were more inclined to convert. As a consequence of proselytization friction with the public was on the rise. In fact, both the Russians and British consulates were concerned that the American missionaries were agitating and destabilizing Iranian Christians and encouraging attacks on Kurds (Yeselson 1956: 122). Despite these protests American diplomats discovered they could respond to such incidents from a basis of power and superiority because of the weakness of the Iranian government and intervene on behalf of American citizens regardless of the nature of their activities in the host country.

Throughout this period the predominant problem between Iran and the US was that the American citizens and missionaries were disrespectful of Iranians, their traditions, values, and religion. It is noteworthy that the opposition to the missionaries came from a wide spectrum of Iranians who resented the attempt to convert indigenous Iranians, both Muslims and Christians (Chaldean Catholics and Armenians and Assyrian Orthodox), resulting in doubts in Iranian officials' minds about the intensions of the Protestant missionaries and consequently resulting in a ban on Iranian citizens from participating in foreign-based religious services (De Novo 1963). In general the feelings of Iranians towards US citizens were mixed. On one hand they had animosity towards US citizens due to rudeness of the latter and at the same time they had admiration for the US for several reasons. One source of admiration was based on the US victory against Britain in 1781. The other source of admiration was US advancement in technology, education, industrial production (especially in military production), and innovation. During this era, US policy towards Iranian citizens was to attract Protestants and young and talented craftsmen (Yeselson 1956: 43). In some cases, some of the Iranian immigrants were sent back to Iran to act as missionaries. In fact, copying the example of the American missionaries, other countries such as France, Russia, and Britain sent missionaries into the four corners of Iran. As one might expect, many Iranians were doubtful of their sincerity.

Throughout this period the government of Iran simply complained that the US government was too aggressive in supporting its citizens. In 1904 Benjamin Labaree, the editor of "Rays of Hope" newspaper (and a companion) were killed. The accused was from Kurdish descent (Malek 1350 HS (1972 AD): 52-53) The Kurds constitute a substantial portion of the population in the northwestern part of Iran, especially around the boarder (the current day Iraq and Turkey). The relationship between the Kurds and the missionaries was fairly hostile. On many occasions the US and even British diplomats had urged the Iranian government to take military action against the Kurds to punish them for their resentment and sometimes hostility toward the American Protestant missionaries, as in the above-mentioned case. The relatives of the accused entered into negotiations with the government of Iran and were ready to pay a fine to release the accused. Although the US government seemed to agree with the solution reports surfaced that the accused would be executed. The Kurds attacked the jail, released the accused, and sent him to the Ottoman Empire. In the process four Iranians died. Talks between the Iranian government and the British Consulate (that was protecting the US interest in the region) were not successful and the government of the United States demanded that the Iranian's primary negotiator be expelled from the region to "avoid revolt" in the region. There was an attempt to intimidate the British Consulate. The regional commander of the army was ordered to put on a show of force, but he was unable to mobilize his troops. The American and British government decided to respond forcefully. Meanwhile, the accusers contacted the Russian Consulate to protect them.

As a consequence to the above activities, the Ottoman Empire began meddling in the region to make sure that the Ottoman Empire was not left out. At this time the US government was demanding that 12 people be punished for Labaree's death, and did not agree with the six months delay requested by the Iranian government. One of the leaders of the protesters opposing foreign government involvement by the name of Mojtahed Orumie was arrested and the mayor of the city of Orumiyeh was removed. This is one of the earliest examples of involvement of Shi'a clerics in politics in post Safavid Iran. People revolted and blocked the extradition of the Mojtahed. However, the issue was overshadowed by the news of the Constitutional Revolution in the nearby city of Tabriz and in Tehran. The issue was set aside and no resolution was reached. The main mistake of the US government in this era was that it focused its effort on supporting US citizens in spite of the troubles that they were causing by aggressively promoting religious conversion of Iranian Muslims and Christians.

An import point to consider is the fact that conversion from Islam to other religions is a sin. The religious word for such apostasy in Islam is Ertedad, which in Arabic means "refusal." Any sin in Islam has its own punishment not only in the eternal life, but also in the secular world. Like any progressive legal system the Islamic Laws do change over time, reflecting the sentiment of the population towards the crime and thus the punishment. There has been era when the punishment was death in some Muslim countries. This does not mean that a particular interpretation of the Islamic Laws cannot or have not been non-compromising at times under certain regimes due to different interpretations. However, such outcomes demonstrate the correctness of the earlier statement that the laws are and have been a reflection of their time and people's values. Had the US focused on providing scientific and educational assistance the result could have been much better.

The US diplomatic envoy was not always and unconditionally in support of all US citizens all the time, however. Two examples are Howard Baskerville and W.A. Moore. Baskerville was a teacher in the Presbyterian missionary in Tabriz who sided with the Constitutionalist revolutionaries of Tabriz and helped them organize their military actions. He was not popular in the American Consulate in Tabriz. Apparently, the popular and grassroots movement of Iranian people demanding constitutional rights was not something that the US government wished to support. Undoubtedly, the Constitutional Revolutionary of Iran, the first such revolution in Asia, was nationalistic and independent from the colonial powers of the era. Baskerville died in a bloody battle in Tabriz in 1909. The US government did not protest his death. The second person was WA Moore, a British news reporter (Greaves 1968).

A brief list of other offensive actions by the United States in Iran includes the capitulation law. Another was the appointing of John Malcolm, in part based on the recommendation of the British government, as the head of the consulate of the United States in Bushehr. John Malcolm was a grain and weapons smuggler in and around the Persian Gulf (Abrahamian 2008; Bonakdarian nd). The new title gave him a cover to continue his illegal activities (Mojani, 1384 HS (2005 AD): 53). For a long time Iranians who complained about Malcolm were treated negatively the US government, but in the end the Americans realized that he was simply filling his own pocket and in fact had a dual British-Iranian citizenship (Mojani,1384 HS (2005 AD): 54). Another source of contention was the act of raising the

American flag on Iranian establishments such as pharmacies and US schools as well as pinning of US flags on the chests of students, which was mandatory at the American School in Orumiyeh (Archives of Foreign Ministry1303 HS Box 9 Dossier 20).

Iranians in the United States

Although it seems that substantial numbers of Iranians were living in the United States around the turn of the twentieth century, very little information is known about them. These were mostly from Azerbaijan in northwest Iran, many of whom were Assyrian and Armenians as well as those that were converted to Christianity by the Presbyterian missionaries. Chaos and the unsafe environment of Iran in those days contributed to migration. American policy of the time was to allow only Iranian Protestants, especially young craftsmen, to immigrate to the United States. This policy caused unrest amongst the Orthodox, Catholic, and Muslim Iranians (Mojani 1384 (2005): 43.). Earlier records indicate that Iranians were concentrated in Michigan, Wisconsin, and Illinois. Most of the Iranians were craftsmen and construction workers.

There are some documents about the success of Iranians in all aspects of life and education in the United States. But, unlike the case of Americans in Iran there is no known document of any criminal act or disturbance by Iranian citizens or migrants in the United States in the nineteenth century or the early twentieth century. Undoubtedly there were criminal activities by Iranians in the United States as the case is with any other group of citizens; however, there is no evidence of intervention of the Iranian government, which was not even able to take care of the citizens within its own boarder let alone in the United States. Apparently, all such criminal cases were handled through the US Judicial system and the Iranian Embassy was not involved. None became political issues as was the case for US citizens in Iran. Part of the reason is that the Iranian came to the United States to enjoy a more prosperous life and in the cases of apostasy to avoid prosecution. On the other hand, US citizens traveled to Iran primarily to correct what they considered to be an incorrect religion and to change social and religious values of the citizens.

In 1911, a member of Iran's diplomatic envoy was given orders to establish an Iranian community in California. Apparently, there was some competition among the cities in the region to attract this group of Iranians because they were craftsmen. Finally, an agreement was reached between Iranian diplomats and the city of San Francisco. Although the city offered to give some free land to the community, the Iranians did not accept the offer because the location of the land was far from railroads or navigable waters. Furthermore, by accepting the land the Iranians would have had to become US citizens. Eventually the Iranians purchased the land they preferred. In summary, the main issue with regard to Iranians in this period was the aggressive US missionary activities, disrespectful behavior of US citizens in Iran, and the US government approach to Iran from an aspect of power instead of a partnership (Mojani,1384 HS (2005 AD): 58). Iranian citizens, and the government of Iran since the 1979 revolution, have been sensitive to the issues of conducting business and establishing relations from a perspective of power by the United States rather than a relationship based on mutual respect. As recent as 2010 this issue has been brought up by the government of Iran, directly or indirectly, in all contacts with the United States.

Early Economic Ties

During the nineteenth century, France (and later, to a lesser extent, other European countries) acted as a counterweight and potential partner to reduce the influence of Russia and Britain in Iran (Daniel 2001, Lorentz 2006).The relationship between Iran and France was terminated during the tenure of Amir Kabir, the prime minster of Mohammad Shah Qajar in 1847.

Toward the end of the nineteenth century some economic ties were developed between Iran and the United States. For example, W. W. Torrence was given a license for 25 years to dig artisan wells to help with agriculture and also bring new technology to help with the growth and modernization of Iranian agriculture. However, as was the case in most so called educational and technical endeavors by Europeans and North Americans, his interest was religious advertisement. Except one well in the yard of the American School in (northern part of) Tehran no other well was ever dug (Nezam Mafi, vol. 1: 51).

During Spencer Pratt's (1866-1891) service at the US embassy in Iran he persuaded the US president to consider trade between the two countries and also served as the representative of the Gatling Gun Company (Daniel 2001). He also obtained the license to operate a power company for 60 years, which he sold to Francis Clercue (Mojani, 1384 HS (2005 AD): 65).

Numerous negotiations and contacts were underway between private and public economic groups. In 1901 the City of Buffalo in the United States invited the Iranian government to send a group to visit a construction expo and inspect the equipment displayed there. Although during this period the United States was moving towards self sufficiency, nevertheless, it was eagerly exploring all possibilities for exports. Numerous agreements and contracts were drawn for imports of weapons, ships, and grain from United States and wool and cotton from Iran. Other trades were picking up as well. For example, there were over 20 large stores in New York that were selling Iranian rugs (Mojani, 1384 HS (2005 AD): 68). There were numerous reports of requests from US legal or individual entities to obtain licenses to establish banks, the right to mine minerals (especially oil), and to build railroads. In 1911 Nabildoleh, the Consular of Iran in Washington DC, contacted the Standard Oil Company of New Jersey to investigate the possibility of oil exploration in Iran. He was also responsible for congregating Iranian craftsmen from around the United States into Southern California (Mojani, 1384 HS (2005

AD): 70). He was also instrumental in securing financial advisors from United States.

Eventually, the United States submitted a series of informational requests to Iranian officials about the rate of exchange based on gold and silver; the state of affairs of minting coins in Iran; detailed information about Iranian markets; marketing means; a list of goods in demand in Iran; medical conditions and production of pharmaceutical products; and Iran's domestic regulations governing employment and trade by Americans in Iran (Archive of Foreign Ministry 1329 HS (1950 AD) Carton 20). There is evidence that the government of Iran provided many and detailed trade related information to the United States, in a sense helping the United States dominate Iran's trade and economy for years to come.

The Role of US Citizens in Iran

Finance

The bankruptcy of the Iranian treasury due to invasion of colonial powers; revolution; excursions by the royal courts to Europe; the royal court's extravagant and lavish life style; and the corrupt activities of government placed the county at the brink of collapse. Finally, under pressure from Majlis, a Frenchman named Bizot was hired to straighten out the nation's treasury department. Shuster (1913) attributes Bizot's failure in his position as being due to his attending too many parties at the British and Russian embassies. Such activity helped Bizot forget why he had come to Iran (Shuster 1913).

Iran was desperately trying to get out of the vicious spiral of decline which was in part caused by colonial powers and in part by inept Shahs and governments, which again were kept in "power" for political reasons. The few nationalist in the government were trying desperately to change this by seeking help from within and without. In pursuit of this objective they also sought financial advisors from Japan.

China was another great nation that was weak in the nineteenth century. China was being taken apart by Russia and Britain, but not at the pace of the same activity in Iran. In 1898, Russia forced China to lease Lu Shun (then Port Arthur), which is located at the tip of the Liaotung Peninsula in Manchuria. Soon after, the Russians occupied the entire peninsula. Between 1891 and 1904 Russians built the Trans-Siberian Railroad (and declined to remove their troops from the region). On the other hand Japan also wanted to exploit the riches of China and also was threatened by the approach of Russia. Japan considered China its own backyard and was doing everything to force the colonial powers of France and Britain, and the new comer Russia, out of China. Japan wished to do the same in China as Britain did in India. (Britain created and maintained buffer zones around India by installing tribal lords and chieftains effectively stopping the advance of Russia in Afghanistan.) In February 1904 Japan attacked the Russian navy and also landed troops in Korea.

Japanese sealed their victories by defeating the Russian navy in Tsushima in May 1905 (Pollack 1905). Based on this victory and other achievements, the Iranians turned to Japan for help. Although Japan was becoming a major global power and could have been helpful, there were no formal relationships between Iran and Japan. This made it logistically more difficult to work with Japan. In addition, Japan's main interest was in Southeast Asia, especially China, Korea, and Vietnam. Japan was also worried about stretching its influence too thin and thus become unable to consolidate and colonize China. Finally, Japan did not feel strong enough to become directly involved in territories where Britain was dominant and would have reacted negatively if Japan began meddling in Iran. Therefore, given ties established earlier, Iran turned to the United States.

Negotiations with the US to provide a team of financial advisors was about to collapse due to the US's reluctance to get involved in Iran's affairs. The US was hesitant to act in direct conflict with the interests of Britain and Russia (Abrahamian 2008, Ghanea Bassiri 2002, Greaves 1968, United States 1911). But suddenly, President Taft intervened and recommended that Morgan Shuster, who had served under him in the Philippines, go to Tehran with a team of 16 experts. The second Majlis authorized hiring Shuster to straighten the treasury. In addition, a military garrison under a Swedish officer was formed to help him in related matters (especially tax collection) (Abrahamian 2008, Ghanea Bassiri 2002, Greaves 1968, United States 1911). This reveals Iran's desperation and also the competition of colonial powers which were competing amongst each other to take a piece of resources of other countries anywhere in the world they could. In May 1911 Shuster and his advisors arrived in Tehran and became the Treasurer Advisor of Persia. When Shuster confiscated the properties of the Shah's brother as payment for taxes Russian troops landed at the Port of Anzaly and demanded an official apology from Iranian government (Greaves 1968). This reveals that he was secretly paid and supported by Russia.

Both Russia and Britain, and later the US, had spies and agents among the Iranian royal family and high ranking officials. As history revealed later, some of the most powerful corrupt people were brought to power by one or the other imperialist during the expansion era of colonial and imperialist powers. Arguably the most notorious of all were Naseraldin Shah of the Qajar Dynasty and Mohammad Reza Shah of the Pahlavi Dynasty. On December 31, 1911 the Russians hung a popular religious leader in Tabriz named Thaghatol Eslam (Abrahamian 2008, Bonakdarian nd, Ghanea Bassiri 2002, Greaves 1968, United States 1911). In March 30, 1912 Russia shelled the Imam Reza's tomb in Mashhad (Greaves 1968). Under pressure from Russia and Britain, Shuster was ousted in December and left Iran by early 1912.

Russia pressured Iran to replace Shuster with the Russian sympathizer Joseph Mornard who was an advisor working in the Iranian customs office at the time (Bonakdarian 2006). Britain advised Iran not to agitate the situation and accept Russian's recommendation to appoint Mornard to the Treasury. However, the British reciprocated the Russian's muscle flexing by moving British-Indian troops into the southern territory of Iran. In early 1912, southern Iran was considered a zone of Britain's influence.

The Russians were acting like occupiers in the northern parts of Iran. They were supporting large land owners and wealthy merchants and they intervened in tax collections. The Russians were also buying villages and farmlands near their zone of influence for a fraction of their value by using intimidation (Galbraith 1989a). On January 24, 1913 they obtained a contract to build a railroad from Tabriz to the border city of Jolfa. This railroad was completed on February 21, 1915 (Abrahamian 2008). The railroad connected Russia to the capital of Azerbaijan which proved instrumental in later invasions of the region. Meanwhile Britain was strengthening its foothold in the south and expanding its newly built Anglo-Persia Oil Company (Abrahamian 2002, Galbraith 1989a, 1989b, Ghanea Bassiri 2002, Greaves 1968a, 1969b, *The Recent Anglo-Persian Convention* 1907, Wilson 2002). Both countries were trying to fortify their positions in their influence zone without upsetting the other. They were each becoming more valuable as an ally to the other in the war against the axis of Germany, Austria, and Italy.

Morgan Shuster's brief stay in Iran as the Treasurer-General of Persia gave the Iranian people and the Majlis much hope and annoyed the Royal Court, Britain, and Russia (Bonakdarian nd, Ghanea Bassiri 2002, Greaves 1968, United States 1911). Recall that the Majlis hired Shuster in spite of opposition by all of the above. There are numerous correspondences regarding his insensitive behavior. There were also rumors that he was Jewish, which did not help his reputation (Greaves 1968). Ironically, he gained respect and trust among Iranians because of British and Russian opposition to his presence. He actually chose the British Major Charles Stokes, a liberal, as the commander of the Gendarme force that was created to help with tax collection. This force had the power to operate in the entire country which was a violation of the Anglo-Russian Convention of 1907. Shuster was hoping to force these two countries to confront each other, but instead they joined forces and opposed the arrangement and eventually ousted Shuster (Abrahamian 2008, Bonakdarian nd, Ghanea Bassiri 2002, Greaves 1968, *The Recent Anglo-Russian Convention* 1907, United States 1911). Based on their own interests, other European powers such as Italy, Germany, France, and Belgium joined the chorus in opposing Shuster (Abrahamian 2008, Bonakdarian nd, Ghanea Bassiri 2002, Greaves 1968, United States 1911).

After the invasion of Port of Anzaly by the Russians, the Iranian government dissolved the Majlis and dismissed Shuster (Bonakdarian nd, Ghanea Bassiri 2002, Greaves 1968, United States 1911). Once again, the United States miscalculated the situation and failed to take advantage of the opportunity to have a strong relationship with Iran. According to Ivanov (Tabrizi 1357 HS (1978): 93-94) Shuster was an agent of Imperialism (Abrahamian 2008, Ghanea Bassiri 2002, Greaves 1968, United States 1911). A suspicion that was heightened by appointment of Shuster as Iran's representative in licenses for the "north oil" and the fact that the documents of the era reveal that he was more concerned with the profits of US companies than Iranian interests.

Further documents reveal that the Consulate of Iran in Washington was also involved in a corporation that was formed together with Shuster to obtain international operating licenses around the world (Archive of Foreign Ministry 1329 HS (1950 AD) Carton 20). On the other hand, Iranians regretted that Shuster was ousted because his reforms and procedures were the only financial improvements of the era and were substantial. Later on, Hosain Ala, Iran's ambassador to the United States, found out that Shuster was interested in helping bring US oil companies to Iran, for a fee. Shuster informed Ambassador Ala that the revenues from the oil in southern Iran would not be adequate to secure US investment. In order to secure the U.S. investments more revenue was needed. Therefore, Shuster suggested combining tobacco, and even the oil from northern parts of the country, into the deal. There is evidence that the US government was involved in the negotiations (Ferrier 1982). In February of 1922 the US consulate in Tehran submitted a copy of a contract between the Standard Oil Company of New Jersey and the Anglo-Persia Oil Company, which was prepared by Shuster. The new company was named Perso-American Petroleum Company. In return the Iranian government received a $1,000,000 loan. The evidence indicates that Sir John Cadman of the British Petroleum Company was not only aware of the contract but also supported it. However, the company was never operational.

Petroleum

General Background
Oil has been flowing above ground for centuries in several parts of Iran such as in Azerbaijan. There are several claims to the origin of the name Azerbaijan but all have some link to "fire." One source attributes the name to Atropates who was the Achaemenian satrap of Media. Atropates remained as the satrap even after Alexander's victory. Atropates means "protected by fire." The word Azar, in today's name of the region also means "fire" in Farsi. The name is also attributed to the fact that at some time in ancient history, the people of the region worshiped fire, which was naturally burning in the mountains due to the flow of "Naft". "Naft" is still the term used to refer to 'petroleum' both in Iran and most of its neighboring countries. Regardless of the origin of the name Azerbaijan, there is no dispute that there has been surface petroleum in the region and that the people were aware of its burning capabilities for several millenniums. There is historical evidence that oil was used for burning as early as 700 BC in Azerbaijan. Fire played a major role in the ancient Iranian religion of Zoroastrianism. Even Marco Polo reports the use of oil for burning in the region.

The earliest documented well, dug by hand, is from Absheron Peninsula dating to the tenth century. However, the territory has long been known as the "land of eternal fire." There is evidence of ancient fire temples. The region also boasts the first offshore oil well at Bibi-Heybat Bay near Baku. The site dates to the early eighteenth century. In the 1820s, a distillation machine to obtain kerosene was invented in Baku. In 1844 an oil well was drilled at Bibi-Heybat, the first such act in the world. Many oil related inventions of the time originated from this region. In contrast, the first US oil well, known as the Drake Well after Edwin Drake, was drilled in 1859 in Pennsylvania. For many years Azerbaijan was the largest producer of oil in the world. Ironically, in the early 1900s, Iran was a buyer of the American oil, which is not surprising because the part of Azerbaijan with an oil industry was separated from Iran by Russia in 1813 by the Gulistan Treaty (Lorentz 2006).

Iran's own quest for production of oil started in 1900 when Kitabchi Khan, an Iranian representative at the Paris Exposition, spoke with some British and French politicians/investors and was ultimately introduced to William Knox D'Arcy. By 1901 an envoy was sent to Tehran for negotiations and a concession for oil drilling was made and signed by the Shah in return for £20,000 in cash, £20,000 in stock and 16% of the profits. Kitabchi Khan acted as the representative of the Iranian government while at the same time receiving £1,000 from D'Arcy. In 1902 drilling started in Chiah Surkh in southwestern Iran. In 1903 it hit some oil and gas. Another well produced oil in 1904. Neither well, however, was producing sufficient amount of oil to be economical. By 1905 the concession was sold to the Concession Syndicate, which switched to another site. In 1908, in a region named Maidan Naftoon or "Petroleum Plain", near an ancient Zoroastrian fire temple—erroneously known as the Mosque of Solomon "Masjid Sulaiman" the first productive oil well in Iran became operational. In 1909, a new company by the name of the Anglo-Persian Oil Company (the name has been changed to British Petroleum) replaced the Concession Syndicate Ltd. (Greaves 1968b). In 1914, the year WWI started, Churchill's government purchased 51% of the company with veto power thus firmly involving the British government in Iran's oil (Greaves 1968b). Later in the 1950s this involvement played a major role in the relationship between Iran and England, Iran's history, and the relationship between Iran and the United States.

The historic events in the northern part of the country are as follows. Naseraldin Shah Qajar (1848-1896) gave an oil concession to Mohammad Valy Khan Tonkaboni in 1896. The concession covered a territory corresponding to today's state of Mazandran located in the southeast of Caspian Sea. In 1916, he sold the concession to a Georgian citizen of Russia named Akaky Mededievitch Khoshtaria. The concession was never ratified by the Majlis.

Khoshtaria took advantage of the Majlis' recess and started an exploration expedition. Authorities in Tehran heard about the expedition only when a message from the Customs Office of the Port of Gaz inquired guidance on the request for exemption from customs for all the supplies and equipment for oil exploration by a foreigner. The Finance Ministry complained to the office of Foreign Ministry (the same as Secretary of State in the United States) about the irregular behavior of the foreigners. The Foreign Ministry, irresponsibly, responded that only the equipment that is actually owned by the holder of the oil concession is exempt from customs. Later, this document was used by foreign countries as evidence of the legitimacy of this un-ratified oil concession.

During WWI (in 1917) the Bolsheviks won the civil war in Russia, condemned many of the imperialist treaties of the Tsars and pulled back from some of the occupied territories. The Soviet Union unilaterally cancelled all imperial-based agreements of Russia and its citizens, which also put an end to the oil exploration activities in the northern part of Iran. The joy of Iranians was short lived due to the report that Khostaria was about to sell his oil concession to England, which gave her another excuse to meddle in Iran's internal affairs. The United States, among others, were also after Khostaria's oil concession.

In 1919, the British and Iranian governments singed a provisional agreement that gave the British control of financial and military affairs of Iran, and the right to explore oil even in the northern parts of Iran. This opened England's hand to increase its dominance in Iran. The fact that this agreement was not ratified by the Majlis, however, gave an opportunity to Soviet Union and to the United States to derail the unchecked power transfer to Britain. In 1919, with pressure from the corrupt prime minister and two of his cabinet members, Iran accepted a "loan" from England. England also "lent" advisors to the army and a majority of the ministries. However, the Majlis, realizing that the uneven shift of resources and power to Britain was not in the best interests of the country, refused to ratify the agreement (Katouzian 1979). By this time the country was in turmoil, augmented by events of WWI, increased influence of Britain, and increased pressure from Russia (by then the Soviet Union) and United States to nullify the substantial gains of Britain. The problems were exacerbated by inflation and unemployment. There were uprisings and protests all over the country. Some of the protests were nationalistic, others enticed by one or the other imperialist powers of the era. Rumors were that a military force of guerilla fighters, supported by the Soviet Union, was ready to march to Tehran in late 1920. Finally, in 1921, Reza Khan Mirpanj took over and ended the Qajar Dynasty and established the Phahlavi Dynasty (Katouzian 1979). Ironically, Reza Khan was trained by the Russians and at the time of the *coup d'état* he was the commander of a Cossack brigade. Britain did not object to the *coup d'état*.

US Iran Oil Relations

The first contact between Iran and the US about oil exploration dates back to the first parliament, right after the Constitutional Revolutions of 1907, during a discussion regarding the exchange of ships for oil. The deal never passed the discussion level. Numerous contacts were made between Iran and United States exploring the possibility of establishing a US oil company in Iran to provide oil for US fuel needs in the Persian Gulf and Indian Ocean. Many alternative financing plans were also considered. However, when negotiations seemed to be producing results and the US was to submit a proposal, Khostaria's concession was revealed and Russia intervened, creating unrest in some areas south of the Caspian Sea. As mentioned earlier, the Russians landed troops at the Port of Anzaly as well. The occupation of the northern regions of Iran by Russia in 1911 decisively put an end to all such discussions with the United States. The next set of contacts began in earnest after the 1919 provisional agreement between Iran and Britain which brought both the United States and the Soviet Union in direct opposition with Britain. Both countries were siding with Iranian government, demanding that the purchase of the un-ratified concession from Khostaria (purchased by Britain) be nullified.

In the 1920 Iran's consulate was invited to attend the annual meeting of the American Petroleum Society. Later, an agreement covering the southern parts of Iran was reached with the Standard Oil Company. Soon the Standard Oil Company realized that it had to deal with local tribes, which were accustomed to bribes paid by Britain. Furthermore, the US was not pleased with the central government of Iran and it also learned that Britain had the exclusive right to all oil pipelines in the region. Therefore, the US was forced to negotiate with the British for access to pipelines or to receive a permit for a different pipeline. In March 18, 1922 The Washington Post revealed a secret negotiation between the Anglo-Persian Oil Company and Standard Oil Company. The news had a negative echo in Iran, which interpreted it as a sign of conspiracy between Standard Oil and Britain (Rubin 1995, Zirinsky 1992). Needless to say the entire affair was disrupted.

Another oil company by the name of Sinclair submitted a proposal for oil exploration in the central and southern parts of the country with the knowledge and cooperation of Britain. Concurrently efforts were underway to somehow incorporate oil exploration in the northern parts of Iran as well. One of the objectives of the United States was to reduce the power of Britain and keep the Soviets out of Iran. The advantage for Britain was that it could keep the Soviets out and possibly benefit from the US relationship with Iran. Therefore, very little opposition was shown. In 1921 the Majlis addressed the issue and ratified the concession to Sinclair with no objections (Rubin 1995). However, after the ratification of the treaty both the Soviet Union and Britain registered formal objections, both of which claimed the right to Khostaria's concession.

The agreement with Sinclair was developed in 11 articles with terms that were more favorable to Iran compared to previous oil concession. As mentioned, after the ratification of the concession both the Soviet Union and Britain voiced their opposition and the Anglo-Persian Oil Company claimed that Iran could not cancel the agreement with the Standard Oil Company (Daniel 2001, Rubin 1995). On the day that the Majlis were supposed to ratify the agreement, there was a major fire in the Majlis and arson was the cause. On July 18, 1924 Major Robert Imbrie, a US diplomat in Tehran, was killed by a religious mob. The New York Times (July 24, 1924) stated that the mob believed he was Baha'i. The New York Herald (September 28, 1924), however, attributed the incident to the rivalry of the Standard Oil and Sinclair Oil companies (Rubin 1995, Zirinsky 1992, Stowell 1924, Turlington 1928). The Soviet news agency ascribed the incident to Britain, and alleged that British Petroleum did not want to lose the deal it had with Standard Oil. Regardless of the cause, the source, or the motives of any agent or conspirator, the net result was that the deal between Iran and Sinclair oil ended. This event and the demonstrations that followed provided ammunition that was needed to declare Martial Law by Reza Khan Mir Panj, which finally ended the Qajar Dynasty and established the Pahlavi Dynasty.

The economic scope of the US-Iranian relationship at this time makes clear the economic interests on both sides. However this is where the similarities end. While the United States sought petroleum and concessions to assist the large US oil companies, which in turn enriched their stockholders, Iranian motivations were different. Iran had sought to maintain sovereignty over its natural resources within the context of its precarious position between the Russians and the British. To this end they sought the help of the US to counter the dominance of Russia and Britain and also to get the best possible deal. These terms were, for the most part, determined by Western powers based upon power asymmetries between strong Western powers and the weak and disjointed Iranian state. Caught up in the race for oil, Iran as a whole and the Iranian state in particular, were in no position to harness their own natural resources and needed assistance, yet this assistance came at a heavy price. Without the means to determine the extent of potential oil revenue, or the means to retrieve the oil, Iran was caught in a classic dependency trap, whereby they were depended upon the generosity of the Western oil companies to provide revenue and technical assistance. Ultimately this form of dependency evolved to exploitation as oil profits were squandered by the elite who used the profits to gain access to Western goods that were beyond the means of the vast majority of the Iranian population.

While wealthy, noble, and elite always existed in Iran the blatant excesses that began during this time caused a shift in perceptions in the general population. Instead of traditional elite from the clergy, nobility, and Bazaar, a new elite emerged, one who was seen as gaining prosperity at the expense of society as a whole. Such an attitude would become institutionalized in later years as Iranian nationalism demanded control, not only over Iranian territory but its natural resources as well. The issue of sovereignty and self-determination for the Iranian people and state became ingrained and has contributed much to the way the selectorate has behaved and how it has used this form of institutional/social memory to gain power.

The Cultural Relationship between Iran and United States

The cultural relationship between the United States and Iran is heavily linked to religious contact between the citizens of the two countries. As mentioned earlier, the first diplomatic liaisons from the United States were concerned with the status of Presbyterian ministries and teachers. It seems that some, possibly a sizable number of Iranians in the United States, also had strong religious preferences. Two groups were present among this group, the Christians of different denominations and the Baha'is. Iranian Christians were, and still are, mostly of the Armenian Apostolic Church, while the Baha'is are the followers of the Bab.

Bab is the title of the first leader of Baha'is. The name in Arabic means the "door." It is short for the Gate to God. Bab was born as Mirza Ali Mohammad in Shiraz, a central city in Iran (Cameron & Danesh 2008, Daniel 2001, Lorentz 2006, Sanasarian 1998). Shiraz is within 25 miles from the ruins of the Persepolis. He revealed his "new religion" to a student named Mulla Hosain in May 1844 and adopted the name Bab. He was exiled to the mountains of Azerbaijan and was executed in July 1850 in Tabriz. Most if not all Baha'is were Muslims. Apostasy is a major sin in Islam and at least at one time there was a Fatwa that killing Baha'is is "permissible" (Mobah in Arabic). Mobah is an act which could be considered a sin but for the benefit of Islam and the Muslim it might be done without a sin. Killing an infidel is an example of a Mobah act, which normally would be a sin but with a Fatwa it would not be considered a sin. Killing in a war is similar in nature. Eventually, his followers recovered his body and buried it in Haifa Israel. Bab claimed that he was a Gate to contact with the 12th Imam of the Shi'a. In fact Bab claimed to be the latest such Gate to the awaited messenger. Due to persecution of the followers of Bab in Iran many followers left. Originally, some settled in Cypress and others in Palestine, which is now in Israel. The first group congregated around Mirza Yahya Sobhc Azal, which is known as Azaly, while the latter group congregated around Mirza Hosain Ali Baha and is known as Baha'i (Cameron & Danesh 2008, Daniel 2001, Lorentz 2006, Sanasarian 1998). The earliest information about Baha'is in the United States is that some had resided in Chicago. The claims of a number of followers cannot be verified accurately and independently. It is also said that they were, and still are rich, influential, and educated people. Such claims cannot be verified independently either. According to Mojani (1384 HS, 2005: 136) many Baha'is, both Iranians and US citizens, resided in Chicago and some even began traveling to Ishqabad, (also known as Ashghabat, which was part of Iran until taken by Russia) and is the home of the first Baha'i temple. Ishqabad is the capital of today's Turkmenistan.

Russians supported Baha'is and allowed them to build their first temple in Ishqabad. The significant numbers of travelers to Ishqabad from the United States and even Russia through Iran was noticed by Iranian authorities. Interestingly, the Embassy of the United States supported the travel, which occurred mostly in the eastern part of Iran. The travel was through Sistan, which later became part of the British controlled zone under the Anglo-Russian Entente. At the time, Sistan was of extreme interest for not only Britain and Russia but also the United States. A part of Iran's Sistan was separated by Britain, which eventually became part of today's Pakistan, when that country declared its independence in 1947. Even today, Pakistan and the United States have extremely close ties, and are working jointly in fight against international terrorism. The US support and encouragement facilitated the spread and establishment of Baha'is in northeastern Iran and a nearby region in Russia. The increased importance of rich Baha'is persuaded the Qajar kings to consider contacting Baha'i investors in Palestine, which facilitated the cultural and political activities of the Baha'is.

One of the cultural activities of Baha'is in Iran was the establishment of a school named Tarbiat. The school was established in 1895 in Tehran. The name Tarbiat means "education" or "pedagogy" in Arabic. Later on, a number of educated and famous people expanded the school. One of those men was Mirza Mortazakhan Momtaz-ol-molk who was the Iranian consul in Washington. In 1910, the school officially linked to the Persian-American Educational Society which was located in Washington D.C. This provided a base for the US to send teachers and build schools in Iran. The cultural relationship between the two countries is mixed with political interests. During this era some cultural societies were created in Iran but all are somehow linked, or accused of being linked, to the Ottoman Empire as well as Baha'i groups in Israel and the United States (Abrahamian 2008, Cameron & Danesh 2008, Lorentz 2006). This link to a small minority (accused of apostasy and sin according to Islamic laws and were loathed by the public) suggests that the United States was supporting a group which was opposed by the majority's rule in Iran and was seen as an opposition group. The resentment and discord became more serious after the Islamic Republic came to power.

The cultural relationship between Iran and the United States had its share of problems. Once again the ignorance of American missionaries and the inexperience of US diplomats on one hand, and Iranian's pride and determination in living their lives their way, is at the heart of the problem. The presence of a weak and inept central government in Iran agitated the tenuous relationships. Although Iranians had admired the US's abilities and technological advances, they did not appreciate being looked down upon.

While Iranians were eager to learn new technology, they were appalled by the idea that people from the new world were telling them their religion was wrong and trying to teach them Christianity. They did not appreciate that the missionaries were generating negative publicity about Iran, and labeling them as barbarians (Majd 2006). In this regard, when the missionaries used the 1911 famine of Hamadan to collect aid, many Iranians were offended. In particular, Nabil al-dowleh (Persia's Charge d'Affaires in Washington D.C.) called them charlatans and thieves. He accused them of trying to line their own pockets in the name of aiding Iranians. For example, in a report about the famine, Dr. Susan I. Moody, an American physician, had claimed that the situation is so dire that Iranians were eating their own children (Majd 2006). Later in 1918, the American School in Iran and some of the religious organizations collected money to help the victims of another famine. Distrust was mounting, and Nabilaldoleh accused the Rockefeller Foundation of using the charity as a cover to obtain political, military, and economic intelligence about Iran. According to Archives of (Iranian) Foreign Ministry (1330 HG, 1912 AD) the Deputy Secretary of State in a conversation with the Iranian chargé d'affaires stated "the visit of the mission is for collecting political-economic intelligence.." and the Iranian diplomat revealed that he had been in touch with Colonel House, special advisor to President Wilson about Iranian affairs. The same source also reported that Baha'is were also active in the discussion with Iranian officials, either independently or through cultural centers as well. They even threatened that if the Iranian government did not cooperate with the missionaries in their activities they will contact the US government and get them involved.

Nabilaldoleh was an interesting character. He used to sell stamps in front of the post office. After the establishment of the Constitutional Monarchy, he started working in the government. Soon afterward he was transferred to the Foreign Ministry, and three years later he was appointed as Iran's consul to Washington. Soon, he demonstrated his diplomatic abilities and was able to present a positive portrait of Iran in the United States. He, his beautiful US wife, and rather large family were topics of many articles in the US media. His wife's picture adorned many of the front pages of news papers. He was a high-ranking Baha'i and also a Freemason, and recipient of Masonries highest honor the 33rd degree. He was closely related to the American-Iran Cultural Society and played a major role in sending Shuster to Iran. Needless to say he was instrumental in providing a foothold for US oil companies in Iran. Eventually, he changed his view about the American-Iran Cultural Society and opposed them. Consequently, he actively opposed the work of the missionaries in the two famine outbreaks mentioned above (Rubin 1995, Stowell 1924, Turlington 1928).

Political maneuvering by the West in attempting to maintain its oil concessions played into the helplessness that many Iranians felt toward the West. The beginnings of Reza Pahlavi's rule was a time of great expectations as many felt that the strong man would clear out the old and corrupt government while standing up to the colonial powers. Hopes were high that a new Iran could emerge that would be stronger, modern, and educated in order to reclaim its political, social, and economic sovereignty. Instead, Iran got nothing more than a dictator, who eventually consolidated the power of the central government and unified the country, restoring some of Iran's lost power.

Bibliography

Abrahamian, E., 1969. The Crowd in the Persian Revolution. Iranian Studies, 2(4), 128-150.

Abrahamian, E., 1979. The Causes of the Constitutional Revolution in Iran. International Journal of Middle East Studies, 10(3), 381-414.

Abrahamian, E., 2008. A History of Modern Iran, NYC: Cambridge University Press.

Avery, P.W. & Simmons, J.B., 1974. Persia on a Cross of Silver, 1880-1890. Middle Eastern Studies, 10(3), 259-286.

Bonakdarian, M., 1991. The Persia Committee and the Constitutional Revolution in Iran. British Journal of Middle Eastern Studies, 18(2), 186-207.

Bonakdarian, M., U.S.-Iranian Relations, 1911-1951. In The United States and the Middle East: Diplomatic and Economic Relations in Perspective. Yale

Council on Middle East Studies, pp. 9-25. Available at: http://128.36.236.77/workpaper/pdfs/MESV3-2.pdf.

Bonakdarian, M., 2006. Britain And the Iranian Constitutional Revolution of 1906-1911: Foreign Policy, Imperialism, And Dissent, Syracuse, NY: Syracuse University Press.

Bostock, F., 1989. State Bank or Agent of Empire? The Imperial Bank of Persia's Loan Policy 1920-23. Iran, 27, 103-113.

Brobst, P.J., 1997. Sir Frederic Goldsmid and the Containment of Persia, 1863-73. Middle Eastern Studies, 33(2), 197-215.

Brockway, T.P., 1941. Britain and the Persian Bubble, 1888-92. The Journal of Modern History, 13(1), 36-47.

Cameron, G. & Danesh, T., 2008. *A Revolution Without Rights?: Women, Kurds, and Baha'is Searching for Equality in Iran.* London: Foreign Policy Center

Daniel, E., 2001. The history of Iran, Westport, CT.: Greenwood Press.

Davis, S., 2006. "A Projected New Trusteeship"? American Internationalism, British Imperialism, and the Reconstruction of Iran, 1938-1947. Diplomacy and Statecraft, 17, 31-72. Available at   http://www.informaworld.com/openurl?genre=article&doi=10.1080/09592290500533429&magic=crossref‖D404A21C5BB053405B1A640AFFD44AE3.

DeNovo, J.A., 1963. American Interests and Policies in the Middle East, 1900-1939, Minneapolis, MN: University of Minnesota Press.

Ferrier, R.W., 1973. The Armenians and the East India Company in Persia in the Seventeenth and Early Eighteenth Centuries. The Economic History Review, 26(1), 38-62. Available at: http://www.blackwell-synergy.com/doi/abs/10.1111/j.1468-0289.1973.tb01924.x.

Ferrier, R.W., 1973. The European Diplomacy of Shah Abbas I and the First Persian Embassy to England. Iran, 11, 75-92.

Ferrier, R.W., 1976. An English View of Persian Trade in 1618: Reports from the Merchants Edward Pettus and Thomas Barker. Journal of the Economic and Social History of the Orient, 19(2), 182. Available at: http://www.jstor.org/stable/3632212?origin=crossref.

Ferrier, R. W. (1982) The History of British Petroleum Company, Vol. 1. P 377. Cambridge University Press.

Ferrier, R., 1986. The Terms and Conditions under which English Trade Was Transacted with Persia. Bulletin of the School of Oriental and African Studies, 49(1), 48-66.

Galbraith, J.S., 1989. Britain and American Railway Promoters in Late Nineteenth Century Persia. Albion: A Quarterly Journal Concerned with British Studies, 21(2), 248. Available at: http://www.jstor.org/stable/4049928?origin=crossref.

Galbraith, J. S., 1989. British Policy on Railways in Persia, 1870-1900. *Middle Eastern Studies*, 25(4), 480-505

Greaves, R.L., 1965. British Policy in Persia, 1892-1903--I. *Bulletin of the School of Oriental and African Studies*, 28(1), 34-60.

Greaves, R.L., 1965. British policy in Persia, 1892-1903—II. *Bulletin of the School of Oriental and African Studies*, 28(02), 284-307.

Greaves, R.L., 1968. Some Aspects of the Anglo-russian Convention and Its Working in Persia, 1907-14--I. Bulletin of the School of Oriental and African Studies, 31(1), 69-91.

Greaves, R. L. 1968. Some Aspects of the Anglo-Russian Convention and Its Workings in Persia, 1907-1914--II. *Bulletin of the School of Oriental and African Studies, University of London, 31*(2), 290-308.

Hart, A.B., 1914. The Monroe Doctrine. The Journal of Race Development, 4(3), 370-373. Available at: http://links.jstor.org/sici?sici=0002-8762(191607)21:42.0.CO;2-B&origin=crossref.

1942. Iran- USSR- Great Britain. The American Journal of International Law, 36(3), 175-179.

Katouzian, H., 1979. Nationalist Trends in Iran, 1921-1926. International Journal of Middle East Studies, 10(4), 533-551.

Ladjevardi, H., 1983. The Origins of U.S. Support for an Autocratic Iran. International Journal of Middle Eastern Studies, 15(2), 225-239.

Lorentz, J.H., 2006. Historical Dictionary of Iran 2 ed., Lanham, MD: Scarecrow Press.

Mahdavi, S., 2005. Shahs, Doctors, Diplomats and Missionaries in 19th Century Iran. British Journal of Middle Eastern Studies, 32(2), 169-191. Available at: http://www.informaworld.com/openurl?genre=article&doi=10.1080/13530190500281432&magic=crossref||D404A21C5BB053405B1A640AFFD44AE3.

Majd, M.G., 2006. Oil and the killing of the American Consul in Tehran, Lanham, MD: University Press of America.

Malek, Rahim Rezazadeh, 1350 HS (1972 AD). Tarikh Ravabet Iran v Mamalek Motahedh Amrica, Tahory.

Metz, H.C., 1989. Iran: A Country Study 4 ed., Washington, D.C.: Library of Congress.

Mojani1, Sied Ali 384 HS (2005AD). Barrasy Monasebat Iran of Amrica (1851-1925) Makaz Asnad v Tarikh Diplomacy.

1999. *Nusrat al-Dawla, Majm'a-i mukatabat, asnad, khatirat-i Firz Mirza Firz (The correspondence, documents, memories of Nusrat al-Dawla)* . Ed. Mansoureh Ettehadia (Nezam Mafi). Tehran

Pollock, A., 1905. The Russo-Japanese War: Its Lessons for Great Britain and the United States. The North American Review, 180(579), 243-248.

Powell, E.A., 1923. By camel and car to the peacock throne, NYC: The Century Co.

Report 284 1329 HG (1911) Archive of Foreign Ministry Carton 20 Dossier 1.

Archives of Iran Foreign Ministry 1331HS (1952 AD)

Archives of Iran Foreign Ministry 1301(1922 AD)

Saleh, J., 1978. *Social Formations in Iran, 750-1914.* Dissertation: University of Massachusetts Amherst

Sanasarian, E., 1998. Babi-bahais, Christians, and Jews in Iran. *Iranian Studies* 31(3&4): 615-624

Seward, G.F., 1912. The Government of the United States and American Foreign Missionaries. The American Journal of International Law, 6(1), 70-85. Available at: http://www.jstor.org/stable/2187397?origin=crossref.

Shuster, W.M., 1912. The strangling of Persia : a record of European diplomacy and oriental intrigue, London: T. Fisher Unwin.

Tabrizi, G.N., 1979. *Technical Education in Iran: Attitudes of Students and Employers*. Thesis: University of Illinois at Urbana-Champaign

1907. The Recent Anglo-Russian Convention. The American Journal of International Law, 1(4), 979-984. Available at: http://www.jstor.org/stable/2186511?origin=crossref.

Thornton, A.P., 1954. British Policy in Persia, 1818-1890-- I. The English Historical Review, 69(273), 554-579.

Thornton, A.P., 1955. British Policy in Persia, 1858-1890. The English Historical Review, 70(274), 55-71.

Turlington, E., 1928. The Financial Independence of Persia. Foreign Affairs, 6(4), 658-667.

United States, 1903. Foreign Relations of the United States- 1903. Prospects.

United States, 1911. Papers Relating to Foreign Relations of the United States, 1911. Bliss.

State, U.D., 1953. First Progress Report on Paragraph 5-a of NSC 136/1, "U.S. Policy Regarding the Present Situation in Iran", Washington, D.C.: State Department.

Volodarsky, M., 1983. Persia and the Great Powers, 1856-1869. Middle Eastern Studies, 19(1), 75-92.

Volodarsky, M., 1985. Persia's foreign policy between the two Herat crises, 1831-56. Middle Eastern Studies, 21(2), 111-151. Available at: http://www.informaworld.com/openurl?genre=article&doi=10.1080/00263208508700620&magic=crossref||D404A21C5BB053405B1A640AFFD44AE3.

Wilson, K.M., 2002. Creative Accounting: The Place of Loans to Persia in the Commencement of the Negotiation of the Anglo-Russian Convention of 1907. Middle Eastern Studies, 38(2), 35-82.

Yeselson, A., 1956. United States-Persian diplomatic relations, 1883-1921, New Brunswick, NJ: Rutgers University Press.

Zirinsky, M.P., 1993. Render Therefore unto Caesar the Things Which Are Caesar's: American Presbyterian Educators and Reza Shah. Iranian Studies, 26(3), 337-356.

Chapter 3
End of Qajar and the beginning of Pahlavi Dynasty (1914-1925)

The rise of Reza Pahlavi indicates how new elite began to take power in Iran and changed the shape of governance. The corruption of the Qajar's was highlighted by the initial austerity and strong rule of Reza Pahlavi.

Ahmad Shah was the last Qajar king. He succeeded his father on July 16, 1909. Since he was born on January 21, 1898 he was still a minor and could not be crowned as King. He was crowned on July 14, 1914, just three weeks before WWI. He immediately was sent to Europe for vacation. Upon the onset of WWI, Iran declared its neutrality. When the third Majlis resumed on November 1, 1914 it appointed Mostofi al-Mamalek as the prime minister. He was known to support neutrality. Parliamentary members of the Democrat Party favored joining Germany, and the Moderate party believed that since Iran was occupied by Britain and Russia it was better to declare neutrality.

At the time there were over 8,000 Russian troops in addition to the Russian guards protecting Russian embassies and consulates. This made the Ottomans nervous, and they invaded Iran on October 1, 1914. The Ottomans also warned that if Iran did not expel the Russian army it would occupy Azerbaijan. Consequently, Iran asked Britain and Russia to respect Iran's neutrality and withdraw their troops. The response was an increase in troops and on December 1, two Russian divisions entered Iran and advanced towards Ottoman territory. On January 30, 1915 Ottomans occupied Tabriz. On April 1916, they defeated the British army during the Siege of Kut. These events created a ray of hope for Iranians. They were dreaming of freeing themselves from the imperialist powers of Russia and Britain. At the time, Iran had a military power consisting of 8,000 in the Ghazag Brigade (under Russian officers) and 7,000 Gendarmes under the command of Swedish officers. The power gradually was tilting in favor of the Entente Powers. Most parts of Iran were cleared of German and other Central Power soldiers. On February 1917, British troops defeated the Ottomans in Kut and retook the city. In April of the same year, Russian and British forces reached Bakhtaran (Kermanshah). On March 15, Czar Nicolas II abdicated his crown. With turmoil in the Russian ranks the Ottomans managed to recapture the city of Bakhtaran, the state of Kurdistan, and the city of Tabriz.

Possibly no other country celebrated the collapse of the Tsars more than Iranians. The hope was that centuries of humiliation, defeat, and loss of territory was over. The sense of disappointment is imaginable as the provisional government was establish in Petrograd, and its Secretary of State informed the Iranian government that they will continue cooperating with Britain and that their policy towards Iran had not changed. However, the October revolution settled the extent of Russian power in Iran and they started evacuating the country (Daniel 2001, Abrahamian 2008) Britain immediately moved in to fill the void and ensure the Bolsheviks did not gain a foothold. The Soviets signed a peace treaty with the countries of the Central Powers, surrendered the territories that they had lost in the war, paid war retribution, and agreed to evacuate Iranian territories, provided that Ottoman Empire also evacuates from the regions they had captured during the WWI.

In response to troop withdrawal by the Soviet Union and the Ottoman Empire, Britain began agitating the people of the Caucasus hoping to entice them to revolt and to seek independence (Bropst 1997). This was in line with long-standing British policy of creating a buffer between potential danger and its interests (Greaves 1986, Greaves 1991, Ingram 1973). In June 1918, it succeeded to replace the commander of Baku with an anti-communist group, and a Muslim Nationalist army was formed with the help of a British general (Zirinsky 1992). The Ottoman response was to attack and capture Baku on 14 September 1918 and create a new, Ottoman-friendly government: Azerbaijan. On October 20, of the same year, the Ottomans ran out of resources and requested a truce. The truce meant the Ottoman evacuation of Baku, and British relinquishment of related territory. On May 26, Georgia, and on May 28, Azerbaijan and Armenia declared their independence. Once again, Britain created a safety belt around its old foe. At the same time, General Sir Wilfred Malison began the task of occupying the Imperial Russian territories of Central Asia.

On December 19, 1917, the Soviet Union cancelled all the privileges obtained by the Tsars from all its neighbors and adversaries. On January 14, 1918, the Soviets specifically canceled privileges, special guarantees, advantages, extra-territorial rights, Consular Courts, concessions, and especially the Capitulation treaty. However, General Baratov and the commander of the Caucasus front refused to obey the Soviet command. They refused to evacuate from Iran. Furthermore, they tried to link up with Britain to overthrow the Soviet Union. Even after the arrival of the Soviet attaché to Tehran, Von Etter, the previous diplomat, refused to surrender the embassy. Since the outcome of the civil war in Russia was uncertain, Iran decided to wait and see. Finally, on July 26, 1918 the government of Iran under the leadership of the Prime Minister Samsam al Sultana (who came to power in May of that year) passed a bill that canceled Capitulation because it was part of the Turkmenchay Treaty with the Russian Tsar, which no longer existed. Interestingly, this infuriated Great Britain, which refused to accept it. In response Samsam al Sultana stated that Britain had violated Iran's neutrality and declared the Southern Police (British) to be a foreign force. The Southern Police were to be dismantled as soon as possible and British forces to evacuate from the country as soon as possible. The result was that Samsam- al- Sultana was forced out of office. Therefore, Britain was the only real remaining power in the country. The other two remaining world powers, the United States and France, only had a diplomatic presence in Iran. The Soviets were an unknown and weak entity.

In June 1919, the Soviet Union sent its envoy to Iran to ratify the cancellation of Capitulation and other treaties and to surrender the railroad and port facilities that were in their possession to Iran. Vusiiq-ul-Daula, prominent Iranian politician of the era and government minister, who was a paid servant of Britain and was secretly negotiating with Britain, refused to receive the Soviet diplomatic envoy. On August 9, 1919 a secret treaty was signed between Vusiiq-ul-Daula of Iran and Sir Percy Cox, the British Minister in Tehran. According to this agreement Britain would obtain the monopoly in oversight and control of the military and monetary affairs of Iran. In return Iran would receive a loan from Britain, payment of war damages, establishment of a railroad, and establishment of customs tariffs (Bostock 1989). The British immediately lent 2,000,000 Sterling to Vusiiq-ul-Daula, and sent a military and a financial consultant to Tehran, even before the ratification of the treaty by the Majlis. The agreement was issued by British Foreign Secretary Earl Curzon and one of its main components was the Anglo-Persian Oil Company's monopoly access to all Iranian oil fields including the ones in the northern territories. The most damaging aspect of the agreement was that Iran lost its status as a fully independent nation.

Based on this aspect of the agreement and Iran's declaration of neutrality during WWI, Britain barred Iran's presence in the Versailles Peace Treaty. Both France and the United States, as the second and third most powerful nations of the time, voiced their opposition to the agreement. The United States insisted that Iran be allowed to participate in the Versailles Peace Treaty. The Iranian Minister in France issued a communiqué denouncing the Anglo-Persian agreement. In retaliation Vusiiq-ul-Daula appointed another Minister to France and moved the Embassy. However, the French government refused to accept the new Minister. Even Ahmad Shah, who was sent on a tour of Europe by Vusiiq-ul-Daula refused to support the agreement, although the British government threatened to terminate the Qajar Dynasty (Bostock 1989, Zirinsky 1992).

In April 1920 the communists in Azerbaijan had a *coup d'état* and forced anti-revolutionary forces and their British supporters to flee from the Caucuses. In addition, Armenia and Georgia fell into Soviet hands. On May 18, 1920 the Soviets, who had been denouncing foreign occupation by imperialists, landed their troops in port of Anzali forcing British forces to withdraw further south. The Soviets stated that they would withdraw from Iran as soon as anti-revolutionary forces were cleared from the region (Katouzian 1979).

Nevertheless, with assistance from Britain, Iran complained to the League of Nations. While negotiations between Iran, the Soviet Union, and the League of Nations were underway on June 6 it was revealed that Mirza Kochek Khan, and the Soviets, had signed an agreement to create the Gillan Republic of the Soviet Union. In Azarbaijan a group named "Ghiam", which means the uprising, declared the state to be the "land of the free" or Azadistan, which was also propped up and supported by the Soviets (Rubin 1995, Zirinsky 1992, Ghanea-Bassiri 2002, United States 1936). As a result, Britain curbed expansion around the southern border of the Soviet Union.

A few days after the return of Ahmad Shah, the Prime Minster Vusiiq-ul-Daula, was removed. The new Prime Minister Mirza Hasan Khan Pirnia, Moshir-al-Dauleh declared that the enforcement of the Anglo-Persian agreement would be postponed until the Majlis made a decision. He also sent many of the British financial advisors on vacation or assignments abroad. Then he sent an envoy to have direct talks with the Soviets.

Meanwhile, Britain and the Soviet Union started to negotiate. They agreed that the Soviet Union would cease communist and anti-British propaganda in Iran and respect Iran's territorial integrity. Britain would end support for anti-revolutionary forces in the Caucuses and Turkmenistan and withdraw its forces from Central Asia. Finally, they agreed to evacuate from Iran as soon as possible. On October 20, 1920, British Minister Norman, and General Ironside (the commander of British forces in Iran) met with Hasan Pearnia (Moshir al-Dawleh), politician and four times prime minister, demanding that the Russian Ghazak division be surrendered to the British army, or Britain would stop its financial assistance to the division. The Prime Minster instead demanded that Britain pay it back its overdue share of oil revenue. At the end Moshir al-Dawleh was forced to resign and the division began retreating from its positions near the Caspian Sea. The new Prime Minister removed the Russian officers of the Ghazak division, but fearing popular opposition he refused to hand command to British officers. Instead, he appointed an Iranian commander by the name of Reza Khan with the title of Mir Panj, which is roughly the same as lieutenant general.

In retaliation Britain threatened to withdraw its forces from Iran. The Bank Shahi, which was operated by Britain, announced its closing and made it clear that people should withdraw their money. British subjects were told to leave Iran. In mid-January, General Ironside went to Gazvin and met with the commander of Ghazak forces, Reza Khan Mir Panj (Katouzian 1979). They agreed upon a *coup d'état*. Britain agreed to arm the Ghazak and Reza Khan agreed not to remove the Shah, to avoid chaos, which Britain believed promoted the communist cause. On February 21, 1921 Reza Khan entered Tehran and captured different ministries and government buildings. Ahmad Shah surrendered and appointed Said Ziyaddin Tabatabai as Prime Minister (Zirinsky 1992, Daniel 2001, Abrahamian 2008).

Iran kept insisting that British and Soviet troops leave Iran. Britain had no intention of doing so and the Russians were procrastinating in signing the pending treaty; they didn't want to leave Iran in the hands of Britain again. A major domestic threat was the weapons that the Soviets had distributed among people in the northern parts of the country. Iran demanded their collection. In January 1921 the Soviets collected some of the weapons and shipped them back to Baku. However, two weeks later, they realized that Britain had no intention of evacuating its forces. So, they landed 2,000 soldiers in Anzaly and stationed them in Rasht. Eventually, a treaty was signed in February 26, 1921 (Bonakdarian nd, Lobanov-Rostovsky 1948, Tapp 1951, Zirinsky 1992). The Soviets agreed not to intervene in Iran's internal affairs; forgave all debts; revoked all concessions that were obtained by the Tsars; surrendered all the railroads, roads, and port facilities that they had in Iran; revoked capitulation; and also surrendered some villages and small islands in the Caspian Sea. Iran received the right to navigate in that sea as well. In return, Iran agreed not to give the concessions that the Soviets surrendered to other countries. Iran would forbid White Russians from operating in Iran, and Iran would not allow other countries to use Iran's territory as a base for attacking the Soviet Union. Violation of these terms would give the Soviets permission to move its troops into Iran to deal with those forces directly. On March 23, 1921 Britain agreed to cancel the un-ratified agreement of 1919 negotiated with the corrupt Vusiiq-ul-Daula (Abrahamian 2008, Bonakdarian nd). On May 15, 1921 the last British soldier left Iran. All British military and financial advisors left as well, and the Southern Police were abolished. It took the Soviets until September 8 of that year to evacuate their forces.

This period marks one of the most shameful periods of Iran's history. The country was ruled by inept Qajar rulers, was overrun by Britain and Russia, and was constantly battered by Ottomans. Iran lost vast and rich regions of its homeland to others and was reduced to a powerless and declining country. After the British and Soviet troops left Iran, Reza Khan gradually improved his power by depending on the army and strengthening it. His position was only strengthened by the fact that the government and the Prime Ministers were inept, unpopular, or both. He consolidated different military groups into one central army and destroyed small tribal powers around the country and restored peace (Daniel 2001, Zirinsky 1992).

The collapse the Russia and the end of World War I left Britain as the sole domineering power of the region. The secret agreement between Iran and Britain, and Britain's insistence to keep Iran out of the Versailles peace talks did not sit well with the new emerging power, the United States, which protested both. This improved the relationship with Iran and also provided some hope for Iranians that maybe the United States could reduce the damaging role of Britain in Iran.

In this era there was hope that the United States could free Iran from being a pawn in the machinations of the great powers. The failure of the Untied States to live up to its perceived "good guy" image was not helpful for relations between the two nations and deepened the growing suspicion of American motives.

Pahlavi Dynasty

The establishment of the Pahlavi Dynasty marks the beginning of the history of the Iranian-American relationship where relations reached their zenith. The establishment of a more stable government in Iran enabled the Iranian state to begin to exert control over more if its territory and eventually force the major powers to withdraw from Iran, their proxies and paid chieftains having been defeated or surrendered. At the same time a new political order emerged, one that would eventually centralize power in an absolute monarchy that acted more and more arbitrarily to maintain power by any means. This centralization was foreign to Iranian culture since traditional institutions were local and regional. Many Iranians saw such changes, even for administrative efficiency, as a usurpation of local rights and traditional privileges of local elites. Moreover, many of the reforms, especially those of Muhammad Reza Shah, did not address the welfare of the masses save for symbolic measures. The unintended reaction was an alienation of the majority of the population. Eventually these pressures manifested themselves in the 1979 Iranian Revolution.

The United States played a large role in the establishment of the Pahlavi Dynasty as well as its fall (Bill 1988, Bonakdarian nd, Davis 2006, Ghaneabassiri 2002, Ladjevardi 1983, Ronfeldt 1978, Rubin 1995). American interests were political, economic, and security based. The American selectorate needed an ally in the region and both of the Pahlavi Shahs were good candidates. In a like manner many in the American business community felt that a deal with the Iranians could break the hegemony of the British in the international oil market and erode the power of non- American oil companies. Following WWII American petroleum producers were hopeful that Iranian oil fields would become a second Saudi Arabia and that they could exploit the natural resource as they were on the Arabian Peninsula. American military interests were part of the "Two Pillar" policy that sought to protect the Persian Gulf by having both Iran and Saudi Arabia protect the sea lanes (Bill 1988, Gause 1985, Pryor 1978, Ronfeldt 1978). The main failing of this policy was that Iran has always seen the Persian Gulf as the its internal waterway while the Saudi's took advantage of this sensitivity by calling it the "Arabian Gulf to agitate the situation and have a point of contention as needed and also to try to force Iranians limit their agitation in Saudi Arabia. Given the closer cultural and historical nature of Iran to the Persian Gulf, the Iranians understood their role to be more active and substantial than the Saudi's who saw their role in protecting the Persian Gulf more in the manner of a conduit for American forces maintaining the flow of oil. This differing interpretation of the same policy has significant repercussions for politics and security in the region even today (Gause 1985, Ramazani 2010, Ronfeldt 1978).

As mentioned in the last chapter the United States objected to the Anglo-Persian agreement that was secretly signed between Britain and Vusiiq-ul-Daula (Bostock 1989, Marsh 1998, "The Recent Anglo-Persian Convention" 1907, Zirinsky 1992). The Iranians hated the deal and welcomed the objection from US, the only major global power to do so. This objection made the United States popular in Iran. Another reason for American popularity was the major role it played in the victory in World War I. After Reza Khan created a unified military on November 17, 1921 and brought calm and order to the country, the (fourth) Majlis was able to meet for the first time in six years on June 22, 1921 (Abrahamian 2008, Bonakdarian nd, Daniel 2001, Foran 1989, Ghaneabassiri 2002, Rubin 1995, United States 1936, Zirinsky 1992). The new Prime Minister, Ghavam-ul-Saltane, decided to use the northern region's oil to help strengthen government finances. Based on the United States' popularity, its improved global status, and its technical abilities the

Prime Minister secretly began negotiations with Standard Oil (Abrahamian 2008, Bonakdarian nd, Daniel 2001, Rubin 1995, United States 1936, Zirinsky 1992).

On November 20, 1921 the Prime Minister granted an oil concession to Standard Oil giving the company the right to explore oil in five northern states for 50 years (Abrahamian 2008; Rubin 1995; Bonakdarian nd; Daniel 2001; Rubin 1995). To make sure that Britain and the Soviet Union did not intervene, the Majlis ratified the agreement the same day. Immediately, the Soviets objected by claiming that any exploration near their border must be with their approval. Second, they pointed out that the concession to explore oil was given to Khostaria in 1916 (Abrahamian 2008, Bonakdarian nd, Daniel 2001, Lorentz 2006, Rubin 1995, Tapp 1951, "The Recent Anglo-Russian Convention" 1907). Based on the agreement with the Soviet Union that resulted in nullification of that concession, Iran had agreed not to grant the same concession to any other country (Abrahamian 2008, Bonakdarian nd, Daniel 2001, Lorentz 2006, Rubin 1995, Tapp 1951, "The Recent Anglo-Russian Convention" 1907). Britain also objected immediately on the ground of the claim that Khostaria had sold its concession to the Anglo-Persia Oil Company and Iran could not grant the same concession to another country (Abrahamian 2008, Bonakdarian nd, Daniel 2001, Lorentz 2006, Rubin 1995, Tapp 1951, "The Recent Anglo-Russian Convention" 1907). Iran was able to refute the Soviet claim by pointing out that the agreement was not approved by the Majlis, and thus it was not subject to the Soviet-Iran agreement concerning nullification of concessions among other things. Britain's claim was rejected on the grounds that it had no legal rights. However, Britain managed to block the deal by citing an agreement that they obtained a long time ago (Rubin 1995) dealing with oil pipelines through the southern part of Iran. By refusing to give permission to build new pipelines or use the British-owned pipelines to the Standard Oil Company, Britain effectively nullified the new concession. Therefore, the Standard Oil Company had to cease all its operations in northern Iran (Abrahamian 2008, Bonakdarian nd, Daniel 2001, Lorentz 2006, Rubin 1995, Tapp 1951, "The Recent Anglo-Russian Convention" 1907). Eventually, Britain managed to get half of the concession from Standard Oil, which infuriated Iran and resulted in eventual cancellation of the entire (Standard Oil) agreement (Abrahamian 2008, Bonakdarian nd, Daniel 2001, Lorentz 2006, Rubin 1995, Tapp 1951, "The Recent Anglo-Russian Convention" 1907). It is interesting to note that on one hand Britain challenged the legality of the concession and at the same time bought shares in the same concession to expand its operations to the northern parts of the country.

The government of Iran approached the Americans for help with the finance ministry. Iran requested Shuster to lead the team but the United States declined and instead sent Dr. Arthur Millspaugh, who arrived on November 18, 1922 (Abrahamian 2008, Bonakdarian nd, Bostock 1989, Daniel 2001, Ladjevardi 1983, Rubin 1995, United States 1938, Young 1950, Zirinsky 1992). In retaliation, Britain demanded that Iran pay all its debt to Britain at once. Since the government was bankrupt and Britain was pressuring Iran, Ghavam-ul-Saltane, the Finance Minster, had to resign (Abrahamian 2008, Bonakdarian nd, Bostock 1989, Daniel

2001, Ladjevardi 1983, Rubin 1995, United States 1938, Zirinsky 1992). On January 24, 1922 Ahmad Shah traveled to France for medical reasons. He hoped to convince the French government to come to his aid. However, and the Shah returned home empty handed (Abrahamian 2008, Daniel 2001, Sheikh-ol-Islami 1984). His return was around the time of the resignation of Ghavam on January 25, 1922. The new Prime Minster, Mostofy-ul-Mamalek, began negotiating with Sinclair Oil. These negotiations progressed well (Abrahamian 2008, Bonakdarian nd, Bostock 1989, Daniel 2001, Ladjevardi 1983, Rubin 1995, United States 1938, Zirinsky 1992). On May 8, 1923 Britain gave an ultimatum to the Soviet Union to cease negative propaganda against Britain in Iran and Afghanistan, to stop supporting the Communist Party of Iran, and to recall its Minister plenipotentiary in Tehran and Kabul; otherwise Britain would end the trade relationship between the two countries. Britain also dispatched several war vessels to the Persian Gulf. The Soviets eventually complied. These circumstances, and the failure of trade negotiations between the Soviet Union and Iran, ended the cabinet of Mostofy-ul- Mamalek and the government collapsed (Abrahamian 2008, Bonakdarian nd, Bostock 1989, Daniel 2001, Ladjevardi 1983, Rubin 1995, United States 1938, Zirinsky 1992). Consequently, Reza Khan Mir Panj managed to increase his power and appoint more of his followers to sensitive posts.

In September of 1922, in spite of Britain's all out support, the last Ottoman Pasha was ousted by the Turkish parliament. In October of 1923 the Ottoman Empire ended, and Kamal Ataturk became the President of the Republic of Turkey ("Chronology of Iranian History Part 2"). This increased the pressure in Iran to end their monarchy and establish a republic. Soon thereafter, Ahmad Shah appointed Reza Khan as Prime Minister, and on November 2, 1923 Ahmad Shah left Iran for good. On April 1, 1924 Reza Khan declared himself the Shah of Iran and founded the Pahlavi Dynasty (Abrahamian 2008, Daniel 2001, Zirinsky 1992). Reza Khan, who has been consolidating his power as the commander of armed forces, used his power to establish a new monarchy instead of a republic. The Soviet Union immediately began supporting the new government and re-started trade negotiations which lead to an equitable trade agreement on July 3, 1924. According to this treaty the citizens of Iran and the citizens of the Soviet Union had equal rights in the other country, the chief and the deputy of the Soviet Union's trade office in Iran were granted political immunity, and the Soviet Union would not limit transit trade between Iran and Europe (Abrahamian 2008, Daniel 2001, Mamedova 2009). Note that political immunity for a trade officer is a "mild" form of capitulation (Renton 1933).

Just before the Sinclair Oil Company and Iran could finalize the oil concession of the northern territories, on July 18, 1924 Major Robert Imbrie, a US diplomat in Tehran, was killed by a religious mob (Davis 2006, Rubin 1995, Stowell 1924, 1926, United States 1939, 1940). The result was termination of negotiations. More information about this incident is presented in chapter 2.

Britain was almost happy to see a powerful and central government in Iran that could block the spread of communism or the advancement of the Soviet Union towards India. Therefore, it did not take any action when Reza Khan attacked Shaikh Khazal in Khozestan, even though he had a secret accord with Britain dating back to November 1914 (Abrahamian 2008, Bonakdarian nd, "Chronology of Iranian History Part 2", Dadkhah 2001, Daniel 2001, Ferrier 2010, Zirinsky 1992). Furthermore, Britain did not object to dismantlement of the Southern Police, its embassy and consulate guards (Abrahamian 2008, Bonakdarian nd, "Chronology of Iranian History Part 2", Dadkhah 2001, Daniel 2001, Ferrier 2010, Zirinsky 1992). On May 10, 1927 Britain ceased it practice of capitulation in Iran (Abrahamian 2008, Bonakdarian nd, "Chronology of Iranian History Part 2", Dadkhah 2001, Daniel 2001, Ferrier 2010, Zirinsky 1992). Iran declared that after one year no such privileges will be given to foreigners, a promise that was kept for about 36 years until 1963 (Abrahamian 2008, Bonakdarian nd, "Chronology of Iranian History Part 2", Dadkhah 2001, Daniel 2001, Ferrier 2010, Zirinsky 1992). During that year a new justice system was established and legal and penal laws were approved (Abrahamian 2008, "Chronology of Iranian History Part 2 and 3", Daniel 2001). Finally, on May 10, 1928 (exactly 100 years after signing of the Turkmanchay Treaty) capitulation was revoked in Iran, ending a century long era of history considered to be shameful by Iranians (Renton 1933). During this period the services of foreign advisors were curtailed and Dr. Millspaugh's activities were limited. The US advisors refused to accept the limitation and wanted the same unchecked power and authority of colonial and imperialist countries of the past and hence their services were terminated on June 1927 (Abrahamian 2008, Daniel 2001, Young 1950).

All customs agreements that were granted during the nineteenth century were canceled and new agreements were signed under new laws. The only exception was the Customs of the South, which was still in British hands. Revenue from this agreement was applied to the interest and principal of the loans that had been borrowed during the nineteenth century. This exception ended on May 9, 1928 with Iran agreeing to pay all its debt to Britain within five years. The entitlement to print currency was taken from the Shah Bank (which was in fact a British bank) and was given to the National Bank (Meli Bank) (Abrahamian 2008, "Chronology of Iranian History Part 3", Daniel 2001, Ferrier 2010; Karshenas 1999, Kazemi 1985).

Although the Soviet Union was very supportive of the Pahlavi Dynasty, and it ratified the treaty that cancelled all privileges obtained during Tsarist rule, the Soviets kept fishing rights in the Caspian Sea which were given to a Russian on October 1879 by Naser-ul- Din Shah Qajar. The fishing rights were renewed several times (Abrahamian 2008, "Chronology of Iranian History Part 3", Daniel 2001, Ferrier 2010, Kazemi 1985, Mamedova 2009, Nowshirvani 2009, Yarshater 2006). When the latest contract expired in 1925 Iran refused to extend the contract. But, the Soviet Union did not follow the terms of the contract, and did not surrender the facilities. Instead, the Soviet Union asked for arbitration. The arbitrators, lead by Mohammad Ali Foroughi, agreed with the owners of the concession and extended the contract for 15 more years. The Majlis, however, rejected the arbitration and refused to ratify it. On February 1926, the Soviet Union closed all border crossings between the two countries and ended all trade. The ban on trade, especially oil and gasoline, caused significant losses for merchants and the citizens of the northern parts of Iran. For years the relationship between the Soviet Union and Iran oscillated between good and bad, depending on current events of the day. Factors that soured the relationship were the Soviet Union's contribution to insurgency and communist cells in Iran and Iran's coming down hard on convicted activists. A factor that improved the relationship was Iran's refusal to allow anti-Soviet cells to operate from its soil.

The relationship with Britain was not as volatile, but there were considerable issues especially with regard to oil and Iran's share of the Anglo-Persian Oil Company. For example, Britain did not pay any royalties from oil revenues, in violation of the contract, to Iran until 1920. After numerous objections Britain agreed to pay a flat fee of one million Sterling to Iran. From 1921 to 1931, the company paid 24 million Sterling in return for taking Iranian oil (Abrahamian 2008, "Chronology of Iranian History Part 3", Daniel 2001, Ferrier 2010, Kazemi 1985, United States 1947, Yarshater 2006). This was a fraction of what the Anglo-Persian Oil Company had to pay in taxes to Britain. Given the fact that the oil company was paying next to nothing to Iran as compared to its revenue, or the taxes it paid to the British government, the British government was eager and trying to achieve more contracts with Iran. To this end, it would make "gestures of good will" towards Iran such as donating the portion of India-Europe telegraph line that was located in Iran in February 1931. This 'gesture' was hollow due to the invention of wireless communication which made telegraphy obsolete. The other gesture of good will was the relocation of the British Royal Navy from Busher to Bahrain. In 1932 the Anglo-Persia Oil Company notified Iran that its share for the year would be 302,000 Sterling. Iran refused this amount, which was less than a quarter of the previous year, and cancelled Sir Darcy's oil concession. In retaliation Britain sent a threatening letter to Iran on December 2, 1932, submitted a complaint to the League of Nations, and returned several military vessels into Iran's coastal waters. As a result of direct negotiations, on May 29, 1933 a new agreement was signed. Iran's share was increased to 20% of the profits plus 4 Shillings per ton of oil exported, and the field of exploration was limited to 100,000 square kilometers in Khuzestan and Bakhtaran (Kermanshah). The company agreed to expedite technical training of Iranian personnel, and the contract was extended for 60 more years (Abrahamian 2008, "Chronology of Iranian History Part 3", Daniel 2001, Ferrier 2010, Kazemi 1985, United States 1947, Yarshater 2006).

The formation of the Pahlavi dynasty demonstrates how a centralized government could consolidate the various groups in Iran who had various interests. The major hindrance to Reza Shah's rule was the situation Iran was forced into by foreign oil companies, namely those of the Great Britain. The push and pull of international and domestic politics placed Reza in a precarious political position versus the various powerful social groups in Iranian Society.

Relations with other Powerful Nations

On January 1937, the French newspaper Excelsior printed a strongly worded criticism of Iran and Reza Shah. In retaliation, Iran cut its diplomatic relations with France, and for two and a half years the two countries remained without diplomatic ties. Finally, France sent an envoy during the wedding ceremony of the crown prince and apologized for the incident, which reestablished diplomatic relations between the two countries. Relations with the United States soured because of the arrest of the Minister plenipotentiary of Iran Jallal Gafar in rural Maryland (Elkton) on a traffic violation (Mahdavy1387 Iranian calendar (2008): 396). There, he was identified and released. Iran requested an investigation, but at the end the United States refused to apologize. The press in the United States accused the Iranian embassy and thus the Minister Plenipotentiary of smuggling, however, this was never proven. Apparently an incident took place and the U.S. government used it to show its unhappiness with Iran's close relationship with Germany. Therefore, in March 1936 Iran severed diplomatic ties with the United States ("Chronology of Iranian History Part 3", United States 1953). Towards the end of 1938, the United States officially apologized, and in the January of the following year, diplomatic relationships were established once again.

Trade between the two countries also improved, and in 1940, US equipment was used to establish the first radio news broadcast in Iran. However, some believe that Reza Shah's strong reaction and uncompromising stance offended US officials, which would explain the United States' reluctance to intervene when the Soviet Union and England invaded Iran in 1941, as part of their World War II strategy.

Between the two World Wars, Iran's relationship with Germany improved substantially. The Germans played a major role in attempts to industrialize Iran. In 1927, Junkers Airline Company obtained the airmail service concession in Iran ("Chronology of Iranian History Part 3"). In 1928 the concession to build parts of the northern territory's railroad was given to German contractors, and in 1930 the operations of the National Bank was given to German financiers. When Hitler came to power, the relationship between the two nations improved even more. Germans pointed out that both Germans and Iranians are Arians and should unite to fight against communism and imperialism. In November 1935 Doctor Schacht the economic minster of Germany visited Iran and a series of talks concerning expansion of trade between the two countries took place. In the following month, an agreement was reached which resulted in rapid expansion of trade between the two countries. In less than five years Germany's exports to Iran increased fivefold, and Germany became the largest buyer of Iran's raw materials. Hundreds of German technicians and experts flooded Iran's fledgling industrial sector. In 1938, a shipping line between Khorramshahr and Hamburg was established and Lufthansa began regular flights between Tehran and Berlin. Italy also helped Iran with the timely delivery of purchased ships and the training of navy officers.

Iran declared its neutrality in World War II which started in September 1939. However, England confiscated some ships that were carrying German equipment for a smelting factory in Iran. Germany approached the Soviet Union, which had signed a peace treaty with Germany, to permit transit of goods between Germany and Iran. With the Soviet Union's permission, the flow of goods and services as well as German advisors continued. On May 30, 1941 British and Indian forces landed in Basra and defeated Iraqi forces and propped up a sympathetic government in Iraq. On June 8, British and French forces captured Syria and Lebanon. The only country in the region that was not part of one or the other fighting nations was Iran. However, after Germany's attack on the Soviet Union in June 22, 1941, all that changed. In order not to provoke England and the Soviet Union, Reza Shah replaced Prime Minister Matin Daftari, who was educated in Germany, with Ali Mansur who was a conservative. He banned German propaganda literature in Iran and limited the activities of Germans in Iran. Nevertheless on July 18, both England and the Soviet Union simultaneously submitted notes to Iran declaring their uneasiness about the presence of Germans in Iran, and demanded their expulsions (Abrahamian 2008, Bonakdarian nd, Dadkhah 2001, Daniel 2001, Davis 2006, Wright 1942). When Iran refused, England tried to convince Reza Shah that it would be in the best interest of Iran to avoid an attack by the allies provided that, in addition to the expulsion of the Germans, Iran were to give the safe passage of ammunition and access to its cross country railroad system from Persian Gulf to the Soviet Union border, which was the shortest route (Abrahamian 2008, Bonakdarian nd, Dadkhah 2001, Daniel 2001, Davis 2006, Wright 1942). Reza Shah refused to respond, hoping that Germans would advance rapidly enough to save Iran from another invasion by England and the Soviet Union.

On one hand, Hitler had been reminding Reza Shah that Germany was advancing rapidly and promised German troops would reach Iran soon and would protect her against England and the Soviet Union. On the other hand, England and the Soviet Union were pressuring Iran to open the supply line or face invasion. On August 25, 1941 England and Soviet Union forces attacked Iran from the south and the north, respectively (Abrahamian 2008, Bonakdarian nd, Dadkhah 2001, Daniel 2001, Davis 2006). The Soviets advanced into all northern states and did not stop until they either captured the state capital or proceeded to strategic supply lines, all past the respective state capitals. Soviet planes also bombed Tabriz, Rasht, Uromieh (Rezaieh), Qazvin, and the suburbs of Tehran. They also dropped propaganda leaflets against Reza Shah and Germany. England, which had amassed huge forces in Basra and the southern parts of Iraq as well as in the Persian Gulf, attacked from the west and southwest. British troops captured the oil fields of Khuzestan and Bakhtaran (Kermanshah) and took control of the ports and railroad stations of the south. The British navy destroyed Iran's navy, which was stationed in Khoramshahr (Abrahamian 2008, Bonakdarian nd, Dadkhah 2001, Daniel 2001, Davis 2006). Five German and three Italian ships that were offloading cargo at Bandar Khomeini (Bandar Shahpour) were attacked. Four were captured but the captain of the fifth ship managed to sink it. However, they failed to block the mouth of the Shatal Arab River by sinking their vessels and thus failed to block the flow of oil.

Within three days Iran surrendered. The allies demanded that all German citizens, except for Embassy personnel, leave Iran; that Iran facilitate shipments of ammunition, equipment, and supplies to the Soviet Union; and that Iran cease all hostilities and resistance. England agreed to continue payment of Iran's share of the oil revenue and the Soviets agreed to do the same for the fishing company in the Caspian Sea. Before Iran could respond, the allies changed the terms and demanded that all German citizens, which were gathered at the German embassy, be handed over to the allies (Abrahamian 2008, Bonakdarian nd, Dadkhah 2001, Daniel 2001, Davis 2006). Reza Shah was reluctant to take hostile actions against the Germans and refused to agree. On September 10 the allies gave a 48-hour ultimatum to surrender German citizens and to close the embassies of Germany, Italy, Romania, and Poland or the allies would occupy Tehran. Reza Shah procrastinated and did not respond. Therefore, on September 16, the Soviets from the north and England from the south began advancing towards Tehran. Reza Shah resigned from the throne in favor of his son (Bonakdarian nd, Dadkhah 2001, Daniel 2001, Davis 2006).

England was eager to sign a cooperation agreement between Iran, the Soviet Union, and England to make the occupation look like a friendly stay during the war with a provision that the allied forces would leave Iran as soon as the war was over. They knew that after the war it would be easier for the Soviet Union to keep and supply its troops in Iran than for England. Iranians, who were furious over the occupation and feared that Germans might win the war after all, refused to oblige. The fact that the German army had made it to Moscow did not help either. On December 5, 1941 the Germans were forced to withdraw from Moscow and on December 7 the Japanese attacked Pearl Harbor which officially entered the United States into the war. Iranians realized that the winds had shifted and agreed to sign a treaty on January 29, 1942. As a gesture of good will the occupying forces vacated Tehran.

Transportation of ammunition and supplies from the Persian Gulf to the Soviet Union was delegated to the United States, which was supplying the equipment in the first place. The US sent 28,000 soldiers and technicians to Iran. The army stayed at the (then) suburb of Tehran until the end of the war. Once again the United States began having a major role in Iran, which eventually made the United States the dominant force in Iran and the region. The United States's influence lasted until 1979 when the Islamic Republic of Iran ousted the Pahlavi Dynasty. The allies were forcing Iran to declare war on Germany or be left out of peace negotiations, as happened in World War I (Mahdavi, 1387 (2008): 413). To avoid the previous experience after World War I, Iran declared war on Germany on September 9, 1943. Later on February 28, 1945 Iran also declared war on Japan (Bonakdarian nd, Dadkhah 2001, Davis 2006, "Iran-USSR-Great Britain" 1942, United States 1955, 1960, Yegorova 1996).

Thus, prior to World War II, Iran was courted by and actually made some overtones to the great powers of the day. Nations like Germany, which have been defeated earlier and had either lost or never gained colonies or countries of interest were eager to embrace Tehran. Germany's interest in Iran was augmented by the fact that the two major players in Iran were also the two most powerful opponents in the war against Germany. The underlying theme of Iranian subservience to outside powers can be seen during this era as well. The frustration felt by all sections of Iranian society towards other nations coalesces into a greater political activism of the various Majlis's, and growing resentment among powerful social groups such as the Bazaries.

Iran during World War II

The estimated population of Iran around the time of the Allied Forces invasion was 15,000,000. The country was poor by any standard and did not have any real sources of revenue. Although its oil had been extracted commercially for decades, the proceeds were limited in part due to the practices of the Anglo-Persia Oil Company, including the outright refusal to pay royalties at times. The influx of thousands of British, Soviet, and US troops created competition for food and other resources of the country, the consequences of which created massive shortages of everything. The need for currency for Allied purchases complicated the shortages and resulted in very high inflation (Abrahamian 2001, Dadkhah 2001, Daniel 2001). Naturally, the black market was thriving.

Under these conditions Qavam-ul-Saltaneh returned from exile and became the new Prime Minister. Soon Wendell Willkie, special envoy of the President of the United States, visited with the Prime Minister and stated that the United States was ready to help Iran. New US advisors from finance, economics, police, gendarmerie, health, agriculture, oil, and military flooded Iran. The most important group of advisors was the financial advisors under Doctor Millspaugh, who had served in the same capacity from 1922 to 1927. He arrived with a 35-member delegation in January 1943 and was appointed as the head of finance with vast powers to control finance, economics, the treasury, transportation, distribution of food, and price stability. However, it turned out that his true mission was to secure food, rations, and supplies for the Allies. Thus, he failed to secure Iran's trust. Furthermore, the Soviets did not allow him in the northern regions and England did everything in its power to limit his authority (Bodaghi, 1387(2008): 24). All these were green lights for the left and the right newspapers as well as the members of Majlis to attack him. Consequently, in January 1945 the Majlis curbed his authorities. Once again Millspaugh refused to comply and left the country with his advisors.

The Fourteen Majlis

The occupation of Iran by the Allied Forces, and the exile of Reza Shah ended his dictatorship. All newspapers were free to write whatever they wanted, all prisoners (political as well as criminals) were forgiven, and all limits and bans on political parties were abolished. All of this occurred while the country was occupied and the world was at the peak of the greatest world war ever (Abrahamian 2008, "Chronology of Iranian History Part 3" nd, Daniel 2001, Ferrier 2010). The Soviet Union and England, as well as other powers took advantage of the situation and either established or expanded their puppet parties and cronies. After the thirteenth Majlis adjourned, a great campaign for the seats in Majlis began. Although the Tudeh Party (the communist party with a great base in Azerbaijan and Gilan with close ties to the Soviet Union) was well organized and had popular basis, failed to get as many seats as expected (Young 1950). In addition Said Jafar Pishevary, a powerful Tudeh representative from Tabriz, was not approved by the Majlis and was dismissed. Consequently, the Tudeh Party took a hard line in Majlis. Consequently, Prime Minister Sohaily was forced to resign and the Foreign Minister (Secretary of State), Saed Maraghe'I, was elected as his replacement on March 28, 1944. Saed was a conservative, but had served in the Soviet Union for many years and was accepted by the Soviets (Abrahamian 2008, "Chronology of Iranian History Part 3" nd, Daniel 2001, Mamedova 2009).

In February 1944, a group representing Socony Vacuum and Sinclair Oil from the United States and another group representing Royal Dutch Shell (a British oil company) requested concessions to explore for oil in Baluchistan in southeast of Iran ("Chronology of Iranian History Part 3" nd, Ferrier 2010, Kazemi 1985). The frustration of the Sinclair Oil Company has been studied in detail earlier. As explained, they decided not to have anything to do with the northern oil. Since the southern part of the country was all in the hands of British companies, Sinclair decided to try its luck in Baluchistan. The interesting thing is how Royal Dutch Shell and England found out about the planned approach and put their oil exploration offer on the table at the same time ("Chronology of Iranian History Part 3" nd, Ferrier 2010, Sheikholeslami 2010). Although entire affairs of both proposals were kept secret, the Soviets found out about it and on September 1944 the Soviets sent their own delegation to begin their efforts to secure oil concessions in Iran once again. Note that all of these negotiations were taking place during World War II while Iran was under occupation (Abrahamian 2008, "Chronology of Iranian History Part 3" nd, Daniel 2001, Mamedova 2009). Under pressure from the Soviet Union, Saed rejected all three proposals on October 18. However, with escalation of demonstrations against him, Saed was forced to resign. Without a Prime Minster and a government, the Majlis was left alone to bear the brunt of three powerful nations all by itself (Abrahamian 2008, "Chronology of Iranian History Part 3" nd, Daniel 2001, Ferrier 2010, Sheikholeslami 2010). It finally took the advice of Doctor Mosaddeq and passed a law stating that as long as the occupying forces remain in Iran no oil concession can be given and any minister or prime minister that violates this law will be prosecuted (Abrahamian 2008, "Chronology of Iranian History Part 3" nd, Daniel 2001, Ferrier 2010, Sheikholeslami 2010).

The defeat of Germany ruled out the possibility of using that country to curb the activities of the Soviet Union and England. France was weakened considerably and was not in any position to play a dominant role in world diplomacy. Naturally, Iran turned to the United States to play that role.

The Soviets and Azerbaijan

As soon as World War II ended Iran reminded the three occupying forces that according to their agreement, which England insisted to pass, they must leave Iran. The US shifted some of its forces to Japan on June 10, 1945 and reduced their forces in Iran to about 6,000. This raised public opinion toward the United States. When on September 2, 1945 Japan surrendered and World War II officially ended, Iran once again requested that Allied Forces leave Iran within six months according to the signed agreement. The three occupying forces agreed and published a *communiqué* to that effect. Within 24 hours of the *communiqué* a group of armed people began an armed uprising in cities of Azerbaijan and, with the support of the Red Army, captured some of the government buildings. Immediately, they declared the establishment of the Democratic Faction of Azerbaijan under the leadership of Said Jafar Pishevary, who was rejected from the 14th Majlis ("Chronology of Iranian History Part 3", Haqsenas 2010, Mamedova 2009). The Democratic Faction demanded that Azerbaijan become independent, declare Turkey (Azeri) as the official language, not pay taxes to the central government, and establish its own military and treasury. The Tudeh party (the communist party) immediately declared its support (Abrahamian 1970, 1978, Chaqueri 1999, Young 1950). In a short period many of the cities and villages of Azerbaijan were captured by the Democratic Faction.

The cabinet of the Prime Minster Sadr, which was appointed by the majority on July 6, 1945, had many rightwing extremist. Reactionary elements were not supported by the minority. In order to avoid the approval of the cabinet the minority began obstruction and effectively refused to allow the Majlis to convene. Therefore, no action could be taken against the Democratic Fraction and finally on October 21 the prime minster was forced to resign.

All attempts to persuade the Soviets to evacuate Iran failed. The fact that the leftist were consolidating their forces in Azerbaijan, Gilan, and Kurdistan meant that the Soviets were going to stay in Iran to prevent central government attempts to crush those movements and also to assist them to declare their independence and join the Eastern Bloc. Eventually, the United States (the undisputed real winner of the World War II) intervened to curtail the Soviet expansion, but the Soviets responded that the events in the northern and northwestern states are reflections of the peoples' desire to free themselves from the reactionary powers of the central government and had nothing to do with Soviet intervention or meddling. In response, the United States discontinued its troop withdrawal and sent 3,000 soldiers to reinforce the garrison in Tehran's suburb. England sent three divisions to the oil rich regions to reinforce its forces in preparation of possible attack by the Soviets (Abrahamian 1970, 2008, Chaqueri 1999, Davis 2006). On December

14, 1945 Pishevary declared the establishment of the self-ruled government of Azerbaijan.

The secretaries of states for the United States, England, and the Soviet Union met to discuss Iran's situation. The Soviets refused to allow the Prime Minister of Iran to join the talks. The meeting concluded that an observer commission consisting of US, Soviet, and British representatives should convene to oversee Iran's affairs similar to the countries in Eastern Europe and be responsible for evacuation of foreign troops (Daniel 2001, Davis 2006). The minority group in Majlis objected strongly and pointed out that it resembles the agreements of 1907 and 1919 which effectively meant suzerainty (Abrahamian 1978). The government was forced to reject the proposal and the prime minister had to resign on January 20, 1946. The day before his resignation the prime minister filed Iran's official complaints against the Soviet Union's occupation with the United Nations. On January 22, Kurdistan's Kumeleh Party (a leftist party) declared the establishment of the Republic of Kurdistan. Iran's complaint to the United Nations was the first case in the newly established international organization and attracted the world's attention. The assembly scheduled to address Iran's complaint on February 28.

On January 26, Qavam-ul-Saltaneh became the new prime minister. He ordered Iran's ambassadors to the Soviet Union and England to contact the Soviets and ask the main purpose of their support for the leftist movements in Iran. When no response was given he requested to meet with Stalin. Stalin demanded that Iran recognize Azerbaijan's independence, give oil concession of the northern states to the Soviets, and to coordinate its foreign policy with the Soviets (Abrahamian 2008, Daniel 2001, Davis 2006). Qavam refused, and decided to return home. Stalin softened his stance and suggested a joint company to explore for oil in the northern states. Ghavam agreed to have a friendlier attitude towards the leftist movement in Azerbaijan. On April 4, 1946 an agreement was signed between the two countries. Accordingly, Soviet troops were to withdraw within 45 days, a joint oil company would be created within seven months of ratification of the treaty by the Majlis, and a peaceful solution was to be formed for the study of the situation in Azerbaijan (Young 1950). On June 14, an agreement was signed between the autonomous government of Azerbaijan and the Central Government declaring Turkish the official language of Azerbaijan; the Democratic Assembly became a State Assembly; the governor of Azerbaijan would be recommended by the State Assembly and approved by the Central Government; and Azerbaijan would have its own military and treasury. On August 2, three communist Tudeh leaders were included in Qavam's cabinet, very much in line with the patterns of gradual shift of Eastern Bloc countries toward becoming satellites of the Soviet Union. Soon a group named the Provisional Society of Khuzestan sent a telegram to Qavam and requested the same treatment for Khuzestan as was given to Azerbaijan. The communist party of Tudeh had many followers in the oil industry of the region and was able to set off a demonstration by over 100,000 oil refinery workers (principally from Ahwaz). In demonstrations that erupted 17 people were killed and 150 were wounded in clashes between demonstrators and government forces.

The communists sabotaged the Abadan refinery. In response, England moved three warships to Basra (Abrahamian 1978, 2008, "Chronology of Iranian History Part 3", Daniel 2001).

Early in September a right-wing group named the "Society of Struggle" against Tudeh was established in Shiraz, in southern Iran. On September 23 they demanded three Tudeh members of the cabinet be fired and that the same rights and privileges given to Azerbaijan would be given to the State of Fars. Fars rebels gradually began capturing cities in the state and some of the ports in the Persian Gulf. By replacing the three cabinet members, the situation in Fars calmed down (Abrhamian 2008, "Chronology of Iranian History Part 3", Daniel 2001, Davis 2006).

While Qavam seemed to be leaning towards the east, he was using the United States to expand and equip the military. Once the army seemed to be strong enough, Qavam declared that, in order to guaranty election results, the army must be responsible for law and order in the entire country. The communist party of Tudeh and the Democratic fraction of Azerbaijan objected. On November 24, Iran notified the United Nations and then deployed three divisions towards Azerbaijan. The ability of the Prime Minister to arm loyal divisions in the army using US weapons was quite an accomplishment in light of the fact that army was one of the strongholds of the Tudeh Party. The same day the US ambassador declared that Iran had every right to deploy its military to Azerbaijan and advised foreign powers not to intervene. This was the first time the United States flexed its influence, gained after WWII, to inform other powers, especially the Soviet Union, that in their calculations regarding Iran they have to include the role of the United States. On December 12 the army entered Tabriz and put an end to the leftist government of Azerbaijan. On February 24, 1947, the Kurdistan movement was destroyed as well. Many of the communist party's leaders were arrested in Tehran (Abrhamian 2008, "Chronology of Iranian History Part 3", Daniel 2001, Davis 2006).

The main development in this era was that the selectorate included an openly communist party. As such, attempts were made to destroy the Tudeh yet it would return later. This type of reoccurring left-wing party symbolized a split in the selectorate and society as a whole. It also saw the establishment of a separate Azerbiajani state. Although Azeri state was short lived it serves as a stark reminder that deeply ingrained in the Iranian psyche is the idea that Iranian as a whole value political and cultural independence. While many Azeri's did not espouse the leftist ideology of the Tudeh they did see independence as an overall social good for themselves. While independence was not to be, the idea of independence for the whole of the Iranian nation was cemented in this era. The increasing penetration of outside ideas and customs would serve the purpose of reminding Iranians that colonization could not only be physical but technological, political, and cultural as well.

The Soviet interference in Azerbaijan heightened the nationalistic feeling not only of Azeri's but of Iranians in general. The idea that a great power could slice off a major part of Iran was anathema to Iranians. The inability to prevent abuses by the great powers in World War II weakened the government while strengthening the hand of groups like the nationalists who sought a stronger Iran-an Iran capable of resisting outside interference. This nationalism is also seen in other groups such as the educated class and even the Bazari class. The educated middle class, albeit small, wanted growth and a strong Iran, the Bazari's wanted a secure Iran so they could conduct business unfettered by domestic or international interference. The clergy, who did not want to enter politics, stressed the centrality of an Iranian way of life that formed a foundation for the nationalists. The combined efforts of these groups resulted in an awakening of nationalism and a greater role for political participation. The new political participation by the mass of Iranians was the catalyst that came to logger heads with American interests.

Bibliography

Abrahamian, E., 1970. Communism and Communalism in Iran: The Tudeh and the Firqah-I Dimukrat. International Journal of Middle East Studies, 1(4), 291-316.

Abrahamian, E., 1978. Factionalism in Iran: political groups in the 14th Parliament (1944-46). Middle Eastern Studies, 14(1), 22-55.

Abrahamian, E., 2008. A History of Modern Iran, NYC: Cambridge University Press.

Bill, J.A., 1988. *The Eagle and the Lion: The Tragedy of American-Iranian Relations*, New Haven, CT.: YAle University PRess.

Bonakdarian, M., U.S.-Iranian Relations, 1911-1951. In The United States and the Middle East: Diplomatic and Economic Relations in Perspective. Yale Council on Middle East Studies, pp. 9-25. Available at: http://128.36.236.77/workpaper/pdfs/MESV3-2.pdf.

Bostock, F., 1989. State Bank or Agent of Empire? The Imperial Bank of Persia's Loan Policy 1920-23. Iran, 27, 103-113.

Brobst, P.J., 1997. Sir Frederic Goldsmid and the Containment of Persia, 1863-73. Middle Eastern Studies, 33(2), 197-215.

Chaqueri, C., 1999. Did the Soviets Play a Role in Founding the Tudeh Party in Iran? *Cahiers du monde Russe*, 40(3), 497-528.

2010 "Chronology of Iranian History Part 2", Encyclopedia Iranica Online, 2020, Available at www.iranicaonline.org

2010 "Chronology of Iranian History Part 3", Encyclopedia Iranica Online, 2010, Available at www.iranicaonline.org

Dadkhah, K., 2001. The Iranian Economy during the Second World War: The Devaluation Controversy. *Middle Eastern Studies*, 37(2), 181-198.

Daniel, E., 2001. The history of Iran, Westport, CT.: Greenwood Press.

Davis, S., 2006. "A Projected New Trusteeship"? American Internationalism, British Imperialism, and the Reconstruction of Iran, 1938-1947. Diplomacy and Statecraft, 17, 31-72. Available at: http://www.informaworld.com/openurl?genre=article&doi=10.1080/09592290500533429&magic=crossref||D404A21C5BB053405B1A640AFFD44AE3.

Ferrier, R.W., 2010, "Anglo-Iranian Relations ii. Pahlavi Period", Encyclopedia Iranica Online, 2010, Available at www.iranicaonline.org

Foran, J., 1989. The concept of dependent development as a key to the political economy of Qajar Iran (1800-1925). *Iranian Studies*, 22(2), 5-56.

Gause, F.G., 1985. British and American policies in the Persian Gulf, 1968-1973. *Review of International Studies*, 11(04), 247-273.

Ghaneabassiri, K., 2002. U.S. Foreign Policy and Persia, 1856-1921. Iranian Studies, 35(1/3), 145-175.

Greaves, R.L., 1968. Some Aspects of the Anglo-russian Convention and Its Working in Persia, 1907-14--I. Bulletin of the School of Oriental and African Studies, 31(1), 69-91.

Greaves, R. L. 1968. Some Aspects of the Anglo-Russian Convention and Its Workings in Persia, 1907-1914--II. *Bulletin of the School of Oriental and African Studies, University of London, 31*(2), 290-308.

Greaves, R., 1991. Themes in British policy towards Persia in its relation to Indian frontier defence, 1798-1914. Asian Affairs, 22(1), 35-45. Available at: http://www.informaworld.com/index/781558302.pdf.

Haqshenas, T., 2010, "Communism iii. In Persia after 1953", Encyclopedia Iranica Online, 2010, Available at www.iranicaonline.org

Ingram, E., 1973. An Aspiring Buffer State: Anglo-Persian Relations in the Third Coalition, 1804-1807. The Historical Journal, 16(03), 509-533. Available at: http://www.journals.cambridge.org/abstract_S0018246X00002922.

1942. Iran- USSR- Great Britain. The American Journal of International Law, 36(3), 175-179.

Karshenas, M., & H. Hakimian "Industrialization ii. The Mohammad Reza Shah Period, 1953-79", Encyclopedia Iranica Online, 2004, Available at www.iranicaonline.org

Katouzian, H., 1979. Nationalist Trends in Iran, 1921-1926. International Journal of Middle East Studies, 10(4), 533-551.

Kazemi, F., "Anglo-Persian Oil Company", Encyclopedia Iranica Online, 1985, Available at www.iranicaonline.org

Ladjevardi, H., 1983. The Origins of U.S. Support for an Autocratic Iran. International Journal of Middle Eastern Studies, 15(2), 225-239.

Lobanov-Rostovsky A., 1948, Anglo-Russian Relations through the Centuries, *Russian Review* 7(2), 41-52

Lorentz, J.H., 2006. Historical Dictionary of Iran 2 ed., Lanham, MD: Scarecrow Press.

Mahdavi, Abdolreza Hoshang, 1387 HS (2008 AD)Tarikh Rabet Kharejy Iran, Entesharat Amirkabir.

Mamedova, N.M., 2009 Russia ii. Iranian-Soviet Relations (1917-1991). Encyclopedia Iranica Online, 2010, Available at www.iranicaonline.org

Marsh, S., 1998. The Special Relationship and the Anglo-Iranian Oil Crisis, 1950-4. *Review of International Studies*, 24(4), 529-544.

Nowshirvani, Vahid, "Commerce vii. In the Pahlavi and post-Pahlavi periods", Encyclopedia Iranica Online, 2010, Available at www.iranicaonline.org

Pryor, L.M., 1978. Arms and the Shah. *Foreign Policy*, (31), 56-71.

Ramazani, R.K., 1974. Iran's 'White Revolution': A Study in Political Development. *International Journal of Middle East Studies*, 5(2), 124-139.

Renton, A.W., 1933. The Revolt Against the Capitulatory System. *Journal of Comparative Legislation and International Law, Third Series*, 15(4), 212-231.

Report 284 1329 HG (19??) Archive of Foreign Ministry Carton 20 Dossier 1.

Ronfeldt, D., 1978. Superclients and Superpowers: Cuba:Soviet Union/ Iran: United States. RAND: Santa Monica, CA.

Rubin, M.A., 1995. Stumbling through the " Open Door ": The U.S. in Persia and the Standard-Sinclair Oil Dispute, 1920-1925. Iranian Studies, 28(3), 203-229.

Sheikholeslami, R., 2010. Administration vii. Pahlavi Period, *Encyclopedia Iranica Online*, 2010, Available at www.iranicaonline.org

Stowell, E.C., 1924. The Imbrie Incident. The American Journal of International Law, 18(4), 768-774. Available at: http://www.jstor.org/stable/2188849?origin=crossref.

Tapp, J., 1951. The Soviet-Persian Treaty of 1921. *The International Law Quarterly*, 4(4), 511-514.

1907. The Recent Anglo-Russian Convention. The American Journal of International Law, 1(4), 979-984. Available at: http://www.jstor.org/stable/2186511?origin=crossref.

United States, 1936. Papers relating to the Foreign Relations of the United States, 1921: Volume II., 44(3).

United States, 1938. Papers Relating to the Foreign Relations of the United States, 1923 Volume II.

United States, 1939. Papers Relating to the Foreign Relations of the United States, 1924 Volume II.

United States, 1940. Papers Relating to the Foreign Relations of the United States, 1925 Volume II.

United States, 1947. Foreign Relations of the United States, 1932 Volume II.

United States, 1953. Foreign Relations of the United States Diplomatic Papers, 1936 Volume III.

United States, 1955. Foreign Relations of the United States, 1945.

United States, 1960. Foreign Relations of the United States, 1945 The Conference of Berlin (The Potsdam Conference).

Wright, R. & Bakhash, S., 1997. The U.S. and Iran: An Offer They Can't Refuse? *Foreign Policy*, 108, 124-137.

Yareshater, E., 2006. Iran ii. Iranian History: Islamic Period, *Encyclopedia Iranica Online*, 2010, Available at www.iranicaonline.org

Yegorova, N.I., 1996. *The "Iran Crisis" of 1945-46: A View from the Russian Archives*, Washington, D.C.
Young, T.C., 1950. The Race between Russia and Reform in Iran. *Foreign Affairs*, 28(2), 278-289.
Zirinsky, M.P., 1992. Imperial Power and Dictatorship : Britain and the Rise of Reza Shah, 1921-1926. International Journal of Middle East Studies, 24(4), 639-663.

Chapter 4
The Beginning of American Influence

Post World War II Iran

US influence in Iran began to increase dramatically following World War II. The war vaulted the United States into a superpower and changed international relations ever since. While the United States did not intend on becoming a colonial power at that time it did feel that it was necessary to fill the vacuum produced by the pullback of Great Britain and other colonial powers. The period from the end of WWII to the middle of the 1960s was the era of independence for colonies and reduction or elimination of colonial power. The struggle for independence, which was based on self-interest, pride, and the struggle to eliminate poverty was a goal of each nation. However, the presence of the Soviet Union with its Marxist orientation, and later the emergence of Communist China, provided ammunitions to these cause, both literally and figuratively.

The beginning of the Cold War changed the relationship between the United States and the Soviet Union. Iran, which once again was caught in the middle between the two superpowers, tried to strike a balance but was always confronted with the stark reality that they held a precarious position. Given the history of Russian intervention in Iran it was natural for Iran to gravitate toward the United States. Reza Shah leaned toward US assistance yet was not fully convinced that the United States was not simply acting as a new colonial power. Of course no sooner Germany was powerful enough before WWII that Reza Shah turned to her for help. The two nations of Iran and Germany soon found many areas of mutual interest. The US selectorate was united in its support of the policy of Containment and thus supported many actions in Iran that precluded the Soviet Union's influence in Tehran. Moreover, the business interests of the US selectorate wished to secure Iranian oil concession and thus supported actions that strengthened the US hand in Iran.

Iran, on the other hand, was developing a selectorate that saw the need for Iranian independence from all major powers. The impetus for this burst of nationalism was the knowledge that to develop economically they would have to harness oil and gas wealth to transform the country. The one who articulated this message most succinctly was Dr. Muhammad Mosaddeq. Mosaddeq and his fight for economic independence and viability embodied all aspects of the selectorate—the business interests, traditional bazarri interests, and the educated middle class who sought constitutional rule. Mosaddeq was a powerful challenger to US interests which were narrowly construed toward oil concessions and containment of the Soviet Union threat. Only secondary to the US selectorate were the aspirations of the Iranian nation as a whole. The following sections demonstrate how these interests played out within the Iranian and US actions leading up to the 1979 Islamic Revolution. These actions set the stage for the present low-level conflict seen today in US-Iranian relations.

The Beginning of Real Influence of the United States in Iran

The United States was the only major power that did not suffer damage to its infrastructure or production capacity during WWII. In fact her production capacity, especially her manufacturing capacity, surged tremendously. Furthermore, the United States demonstrated without any shadow of a doubt that it had become a super power. Even before the end of the war, the United States began flexing its muscles against the Soviet Union, and to a lesser extent against England. For example, it demanded foreign armies leave Iran and when they refused or began to amass more troops the United States brought its forces back. It was in the interest of the United States to "liberate" other countries from the yoke of colonial and imperial powers. This would reduce the power of such countries, especially the Soviet Union and England, to the advantage of the United States.

In Iran, as well as in numerous other countries, the United States was viewed as the savior of the country. After the problems of Azerbaijan and Kurdistan were resolved, Iran was ready to start the 15th Majlis. By this time, the central government had established its supremacy over the entire country, and through the help of the military, managed to fix the outcome of the election in its favor to the point that the Tudeh party or the Nationalist party could not place a single member in the new Majlis (Abrhamian 1970, 2008, Daniel 2001, Davis 2006, Mokhtari 2005). Even the popular Mosadeq was not elected. This was a sign of things to come in Iran under Muhammad Reza Shah. The United States, with all her support, had no objection to election rigging as long as it kept the Communists or others that opposed Western influence out of the government. In August 28, 1947 the Soviet Union reminded Iran that it had fulfilled its obligation and had evacuated its forces from Iran and Iran should take care of its part of the deal and begin the process of establishing the joint oil company with the Soviet Union to explore the northern territory. By now Iran had established its sovereignty over its entire territory and was counting on the United States to support it against the communist neighbor based on the March 12 speech of Truman, in which he had spelled out the United States' plan to defend small nations such as Greece against the expansion of communism (Abrahamian 2008, Amuzegar 1958, Daniel 2001, Young 1950). Therefore, the Iranian government pointed out that the company's future would be in the hands of Majlis, which in addition to having no Nationalist or Communist members was also padded by 30 members of the armed forces.

A contract was signed between Iran and the United States for the purchase of military equipment worth $10 million on 8 June 1947 in Washington. On October 6, another agreement was signed in Tehran that increased both the numbers of

United States military advisors in Iran and the level of their activities and authority (Abrhamian 2008, Amuzegar 1958, Bonakdarian nd, Pryor 1978). In addition, Iran promised not to hire military advisors from any other country. Thus, firmly and completely, Iran and the United States became allies and the relationship between the two countries entered a new phase. Since England had been struggling to keep the Russians/Soviets as far from India and southern Iran as possible for almost two centuries, she did not appreciate the presence of this new superpower in Iranian territory. Thus, it forwarded a message advising Iran not to interrupt negotiations with the Soviet Union and to not firmly end the wartime agreement between the Soviet Union and Iran. On September 15, the Soviets delivered a note to Iran accusing Iran of adopting unfriendly and discriminatory policies toward them. On October 22, a sizable majority of the members of the Majlis voted against the oil exploration agreement with Soviet Union, and forbid signing any oil exploration concession to any foreign country. The Majlis also ordered the government to begin oil exploration directly and if any oil was discovered in the northern region to begin negotiations to sell it to the Soviets. Finally, it ordered the government to demand the fulfillment of the Iranian rights on the southern oil from the Anglo-Persia Oil Company (Abrahamian 2008, Daniel 2001, Young 1950).

On November 20, 1947 the Soviet Union accused Iran of a hostile attitude towards itself and alleged that Iran had been converted to a base for attacking the Soviet Union. Referring to articles 5 and 6 of the 1921 agreement the Soviet Union threatened to sever diplomatic ties and to launch a military offensive against Iran, per the provisions of the agreement. In early 1948 the Soviet Union reduced commercial trade between the two countries, ended boat shuttles between Anzaly and Baku, shutdown parts of the fishing company's operation in Iran, and began a massive propaganda campaign against Iran (Abrahamian 2008, Daniel 2001, Young 1950).

Needless to say, the United States was thrilled and happy to have closed another section of the safety belt around the communist bloc. Arguably, no place in the world had as many communists and as well organized a communist party as Azerbaijan that did not become part of the Eastern Bloc. This was in spite of the fact that the state was actually taken by communist armed forces and had declared its independence from the Central Government.

The end of the 1940s and the early years of the 1950s was a difficult time for Iran. In addition to the familiar tug of war between the Soviet Union (and its predecessor Russia) and England, a new player had appeared on the scene: the United States. At first, this was positive news, because Iranian politicians were thinking they have finally found a powerful "third power" that could stand up to the Soviet Union and England. What made the possibility promising was that the US talked of democracy and human rights, although its actions in other parts of the world (supporting authoritarian governments and even colonies as in the Philippines) did not match the rhetoric.

Another factor that affected the role of the United States in Iran, and thus the relationship between the two, was the global clash of the West and the East. The results of this clash were the creation of communist countries in Eastern Europe and other parts of the world as well as the creation of pro-Western puppet governments wherever the opportunity existed as well as in strategic areas especially around the Soviet Union, China and other socialist/communist countries. Next door to Iran, the Iraqi government had changed substantially and was more sympathetic to the East. A little to the West, Israel and Turkey had become US strongholds and further down the road, Egypt was in the Eastern bloc's camp. Iran was the only country that bordered the Soviet Union that was not communist yet did not receive substantial military or economic aid from the United States.

After World War II, the United States evacuated most of its troops from Iran, while the Soviet Union and England did not. Apparently, Truman gave the Soviets an ultimatum, and as mentioned earlier, the Soviets withdrew their troops. In 1947 Iran and the United States signed a military agreement and the power of the Soviet Union declined in Iran further. There was a period of calm and stability in which the United States spent most of its time and energy in other parts of the world, most notably Greece, Turkey, and Korea. Communist penetration in Azerbaijan and Kurdistan were a wakeup call for the United States as well as for Iran. The United States war trying to replace, declining British influence, in the Middle East and regions that were considered vital in the fight against communism. On many occasions the United States and England cooperated with each other to stop the Soviet Union.

During the late 1940s and early 1950s, many cabinets formed and collapsed in Iran. One solution was to provide military aid to Iran to support the Shah and also block the Soviets. In June 20, 1947, in line with Truman Doctrine, an agreement was signed between Iran and the United States whereby the United States would provide military equipment and munitions. In addition the United States agreed to lend $25 million to Iran. On February 4, 1949 there was an assassination attempt against the Shah and as a result Ayatollah Kashani was exiled to Lebanon (Abrahamian 2008, Amuzegar 1958, Bonakdarian nd, Daniel 2001, Pryor 1978). On October 6 a military accord was signed between Iran and United States extending the presence of the United States military until March 20, 1949. In November the Soviet government submitted a letter of protest pointing to the agreement signed between Iran and the Soviet Union after WW I banning the presence of foreign militaries in Iran. The Soviets stated they considered the presence of the United States military personnel and advisors a violation of the 1921 agreement between the Soviet Union and Iran (Bonakdarian nd, Gavin 1999, Lambton 1957, Mokhtari 2005, Pryor 1978, Ronfeldt 1978).

On February 17, 1948, the Majlis reduced the proposed loan from the United States to $10 million, demonstrating its concern for obligating the government to foreign loans. The Majlis was also anxious to finalize the oil agreement between Iran and England and was demanding a 50-50 partnership, which the Anglo-Iranian Oil Company and England did not accept (Galpern 2002, Gavin 1999, Kazemi 1985, Marsh 1998, Ramazani 2010). Between this period and the *coup d'état* of 1953, several cabinets formed and collapsed. The most notable one was the government of General Razmara, which was a favorite of both England and the United States. In reality, Razmara assisted the Soviets more than either England or the United State. The one crucial thing he did was delay the signing of the oil contract. He also was very much against nationalization of the oil industry in Iran and managed to derail all oil nationalization efforts. Consequently he was assassinated by a member of the Muslim Fedayeen e-Islam group on March 6, 1951 (Abrahamian 1970, 2001, 2008, Daniel 2001, Ruehsen 1993). Interestingly, Razmara had requested a $250 million loan from the United States. It was also known that he was in favor of a military coup, and the dissolution of the Majlis. The United States had considered supporting his military *coup d'état*. Soon, Mosaddeq, who was the chairman of the oil committee in Majlis, was approved as Speaker of the Parliament. On March 15, 1951, under the leadership of Mosaddeq, oil was nationalized. On April 29, 1951 the Shah finally gave in and signed the bill to make it legal. The Iran National Oil Company was formed and took over oil facilities in Khuzestan. British tankers refused to provide receipt for oil they loaded and 4,500 foreign workers resigned collectively, crippling the production and export of oil. The Shah also gave in to Majlis' demand and appointed Mosaddeq as Prime Minister (Abrahamian 2001, 2008, Daniel 2001, Shoamanesh 2009, Zahrani 2002). Later, the Shah decided to replace him with Ardashir Zahedi, but Mosaddeq refused to step down.

1953 Coup d'état

Iran's movement toward self-government and efforts to nationalize the oil industry did not sit well with England, which still had some power, although it had to take a backseat to the United States and was also weaker than the Soviet Union. On one hand, the United States supported nationalization of oil in Iran, which would cut the British hand off from Iran's oil and, for that matter, most other affairs. As noted earlier, oil concessions to British companies had effectively made it impossible for anybody else, including Iranians themselves, to explore, extract, ship, and export Iran's oil, which effectively kept US companies and interest out of Iran. On the other hand, it was in the United States' self interest to have its finger in Iran's oil business, both for economic as well as political reasons (to block the Soviet Union's expansion). In order to secure the enclave of the Soviet Union, the United States needed a more reliable ally in Iran than the independent minded government of the day could or would deliver. The election of a Conservative government in England in 1951, and the Republicans in United States in 1952, paved the road for bringing Iran in line with the wishes and desires of the United States. The United States' embassy played a major role in masterminding the plan to overthrow the nationalist government of Mosaddeq and, in coordinating the activities, saw to it that the government was overthrown. Since England did not have any diplomatic relationship with Iran it was up to the United States to take care of the *coup d'état's* logistical needs on the ground (Abrahamian 2001, Ruehsen 1993, Wilber 1969, Zahrani 2002). The detail of the coup will follow.

Soon after nationalization of oil on April 29, 1951, and the refusal of British tankers to haul the oil, England filed a complaint with the International Tribunal at The Hague and dispatched its naval fleet from the Mediterranean Sea towards the Persian Gulf in an attempt to take Khuzestan by force. Truman sent a special envoy to Tehran for mediation. Upon failure of the talks England deployed its paratroopers to Cyprus to be able to move them to the scene on the ground in time. It also positioned numerous warships around Abadan, the capital of the state of Khuzestan and where a major oil refinery is located (Daniel 2001, Gavin 1999, Marsh 1998, Pirouz 2001, 2008). The United States opposed military action against Iran to avoid Soviet intervention. The unrest and turmoil forced Prime Minister Ghavam out of office on July 21, 1952 and Mosaddeq was elected Prime Minister, again. England and United States drafted a proposal to resolve the oil crisis, which was rejected by Mosaddeq because it contradicted Iran's Oil Nationalization Law (Abrahamian 2008, Daniel 2001, Gavin 1999, Marsh 1998, Pirouz 2001, 2008). Fearing British attack, he severed diplomatic ties with England and ordered all British subjects out of the country to reduce their intelligence activities and capabilities.

The most prominent figure of the *coup d'état* of 1953 was Dr. Mosaddeq, who had a Ph.D. in political science. He published numerous articles in foreign journals in opposition of the 1919 Anglo-Persia agreement. During the 1949 Majlis election, the issue of nationalizing the Anglo-Iranian Oil Company (AIOC) was a main campaign issue. In that year, the Majlis approved the First Development Plan, which was the first step toward economic development. The plan was supposed to be financed by oil revenues. Therefore, to generate sufficient revenues, the Majlis and the Nationalist Movement were trying to renegotiate the oil contract. Mosaddeq became the chairman of the committee that dealt with the issue in the Majlis. In November 1950 the committee rejected the new AIOC proposal on the grounds that it did not include equal partnership between the company and Iran (Abrahamian 2008, Daniel 2001, Kazemi 1985, Ramazani 2010).

According to documents published by the Central Intelligence Agency (CIA) and the State Department in spring of 2000: between November and December 1952, Britain's MI-6 intelligence agency met CIA agents in Washington D.C. to discuss joint operations against the nationalist movement of Iran. The British agents were Christopher Montauk and Monty Woodhouse. The US agents were John Bruce Lockhart, Kermit Roosevelt, and John Wait. In March of 1953, the United States responded positively to British suggestion (Abrahamian 2001, Koch 1998, Risen 2000, Ruehsen 1993, Wilber 1969, Zahrani 2002). In a report by the New York Times, the United States government initially provided one million dollars to the *coup d'état* committee (Risen 2000). By April 1953, contacts, agents, and operators in Iran were identified. They included military personnel and commanders, religious leaders, freemasons, Iranian intellectuals, and United States and British agents. The purpose of the *coup d'état* was to return the Shah back to power. The main targets were Mosaddeq and Ayatollah Kashani, who was the undisputed spiritual leader of the Shi'a at that time. The date was set and the code name Ajax was chosen for the operation (Abrahamian 2001, Koch 1998, Risen 2000, Ruehsen 1993, Wilber 1969, Zahrani 2002). The documents were signed by the director of MI-6 on July 1, followed by the British Secretary of State and Prime Minister on July 1 followed by the CIA Chief, Secretary of State, and President Eisenhower on July 11 (Abrahamian 2001, Koch 1998, Risen 2000, Ruehsen 1993, Wilber 1969, Zahrani 2002). As is evident, British and US hands were both in the *coup d'état*, but England had no embassy in Iran at the time and the United States was making all the contacts. Although England was able to exact revenge for what Mosaddeq did to them, it did not benefit as much as the United States did. As is customary with these kinds of events, the western media, radio, and television orchestrated massive propaganda against Mosaddeq and Ayatollah Kashani (Abrahamian 2001, Koch 1998, Risen 2000, Ruehsen 1993, Saghaye-Biria 2009, Wilber 1969, Zahrani 2002).

The shift of power from London to Washington, the role that the latter played in the *coup d'état* of Iran, the subsequent installation of Mohammad Reza Shah as the new dictator of Iran, were important moves in the Cold War chess game. The Iranian government called the relationship "Positive Nationalistic Policy" (Abrahamian 2008, Bill 1988, Daniel 2001). As agreed, the Shah declared Ardeshir Zahedi (who had studied in Utah in US and in 1957 became the son-in-law of the Shah) as the Prime Minister, several of the ministers were arrested at midnight, and the army was dispatched to notify the Prime Minister of his dismissal. Mosaddeq, who found out about the plot, countered and had the commander of the group arrested. On August 16 the government announced that the *coup* had failed. Subsequently, the Shah fled to Baghdad (Abrahamian 2001, Koch 1998, Love 1953a 1953b, Risen 2000, Ruehsen 1993, Wilber 1969, Zahrani 2002). Opposition to the monarchy, including the communists of Tudeh Party, began taking over government facilities. Consequently some charlatans who were paid by Ardashir Zahedi using US money took to the streets scaring people by beating, injuring, and even killing them. Therefore, they managed to take control of the streets allowing free access to Mosaddeq's house as well as other leaders of the Nationalist Front party, and high-ranking government officials (Abrahamian 2001, Koch 1998, Risen 2000, Ruehsen 1993, Wilber 1969, Zahrani 2002). Eventually, the Royal troops and the thugs managed to turn the situation around, overthrow the government, and arrest Prime Minister Mosaddeq. After the *coup d'état* the Shah returned and Iran became one of the United States's satellites and a pawn in the Cold War against the Soviet Union. This was the seed of abhorrence of Iranians against the United States. Eight years later the expansion of United States dominance in Iran augmented this feeling. After the *coup d'état* in 1953, the United States provided $23.4 million in technical assistance from the Point 4 program of Truman, and another $45 million in grants from the Joint National Security funds to Iran (Amuzegar 1958, Gavin 1999, Pryor 1978, United States 1953).

According to documents from the National Security Agency, several things are clear. First, the most important thing for Iranians in the early 1950s was nationalization of oil. Second, Mosaddeq was the most popular politician of the time. Third, the Shah was neither popular nor relevant, for anybody from Iranians, to the United States, the Soviet Union, or even England. In fact, there were many discussions of a *coup d'état* by general Razmara, which were at least considered by the United States. Fourth, the Soviets were actively seeking territories to increase their sphere of influence and to stand up to the United States' rapid and expanding control of all the territories that belonged to the last generation of colonialists. Fifth, there was a powerful and well organized communist party (Tudeh) as well as many leftist groups and parties in many parts of Iran. Sixth, the Tudeh party had lots of supporters in manufacturing, industrial sectors, and urban areas. Seventh, England would pursue its own interests, especially maintenance of its previous regions of influence and their resources, such as oil. In spite of all these issues, the United States agreed to join England and have a *coup* against Mosaddeq. This decision might have been due to the fact that Mosaddeq would not sell the country but Shah would do anything to stay in power. Furthermore, Mosaddeq was a great opponent of (British) imperialism, which could be extended to the United States; however, the National Security Council's documents indicate that the United States was not necessarily against him (Abrahamian 2001, Koch 1998, Risen 2000, Ruehsen 1993, Wilber 1969, Zahrani 2002). Mosaddeq's anti-imperialism stance was in line with Tudeh Party's communist agenda, thus they supported him, which could have been interpreted as if he agreed with them. The United States believed that the mounting pressure on Mosaddeq, by England and the United States, and Iran's financial crises due to refusal of the West to purchase any oil from Iran, would have forced Mosaddeq to enter into a coalition with the communists. Mosaddeq always supported everybody's right to assemble and to have their own political party, including the Tudeh party, which grew from 5,000 to 30,000 while he was in power. This freedom did not sit well with the United States.

After the *coup d'état,* a major job of the Shah's regime was to destroy the Tudeh Party. Of course, a major concern of the United States was that if it supported nationalization of oil in Iran, it would have been less effective in opposing such movements in countries such as in Venezuela and Saudi Arabia (Abrahamian 2001, Koch 1998, Risen 2000, Ruehsen 1993, United States 1953b, Wilber 1969, Zahrani 2002).

After the *Coup d'état*

The 1953 coup was a watershed in Iranian/US relations. On one hand the United States had a credible and loyal ally in the Shah, who seemed to fit the model of the authoritarian type of leader that supported United States's interests, opposed the Soviets most of the time, and was friendly to global economic interests. His profile was similar to puppet leaders elsewhere in the world propped up by the United States by other major powers before. Much in the same tradition as Latin American authoritarians, the Shah's legitimacy was suspect and his support was purchased and backed up by United States military equipment (Kinsella 1994,

Ladjevardi 1983, Pryor 1978, Ronfeldt 1978, Sanjian 1999). On the other hand, the vast majority of Iranians felt that their aspirations for independence and national determination had been set back. Instead of viewing the United States as a supporter of nationalist movements and constitutionalism the majority of the population came to view the United States as a supporter of the Shah mainly for his anti Tudeh stance rather than his leadership. Many Iranians saw support for the Shah as support for corruption, limited Iranian sovereignty, and imposed social and economic changes in the name of modernization which in reality was nothing more than selling out to shallow US consumerism. As the following section indicates, the roots of revolution were sown in the *coup* and its aftermath. The preferences of the U.S. selectorate were static, wanting an anti-Soviet regime in Iran with increased access to Iranian oil and gas as well as commercial sales of US goods and services, but most importantly military hardware.

A fundamental change took place in the Iranian selectorate whereby groups who were not in opposition to a heavy hand by their leaders were increasingly alienated by the Shah's rule. The traditional commercial class, the Bazarri's, became anti-Shah as they lost their prominence in the economy to large corporations - most foreign owned (Rahnema 1990). The Shi'a clergy began, once again, to become political as during the Constitutional Revolution of the early days of the twentieth Century. They saw an erosion of traditional values of the Iranian society as well as a seemingly unstoppable Westernization that at times was grossly anti-Islamic in orientation. Most importantly the educated middle-class began to lose faith in the regime, as cronyism and nepotism became the *sine qua non* for advancement rather than merit. While the Shah espoused the virtues of merit the reality was much different, with Western educated professionals seeking political voice only to be silenced. The courtiers of the Shah became wealthy at the expense of the middle class while the more traditional lower classes became radicalized. The United States' faith in the Shah seemed blind to the fact that significant opposition was mounting (Abrahamian 2008, Amuzegar 1992, Daniel 2001, Looney 1988, Rahnema 1990). The power of the US selectorate to dictate the United States's side of the relationship is apparent and shortsighted. Unqualified support for the Shah by the United States disillusioned many Iranians who sought peaceful and incremental change toward a conditional monarch with checks and balances. The failure of the United States to see the rise in opposition to the Shah and turning a blind eye to the Shah's authoritarian activities created a situation where change could not be incremental but had to be radical and fundamental. The fundamental change in the Iranian state and society is the legacy of the Shah and the United States' support for his policies. The Iranian Revolution was a reaction to the Shah's political repression. The dissatisfaction with the Shah's regime was extended towards the United States. The Iranian public was dismayed that American love of civil liberties and democracy was not extended to Iran, and that the U.S. supported an anti-democratic regime.

No one would rather live under a dictatorship without any freedom. Putting up with such regimes is temporarily and due to immense power of such governments, backed up by US weaponry, military advisors, and when necessary by US troops. Sooner or later one country or another will find enough strength to stand up and topple the puppet regime.

Iran-US Relations

The new chapter in Iran-US relations was based on oil, Cold War relations with communists, and support of the Pahlavi Dynasty at any cost. Although the United States agreed to provide some financial aid to Iran to make sure that Shah's regime did not collapse, and to insure that communism did not penetrate into the region, it did not want to provide enough aid to solve all Iran's problems because it wanted to end the dispute between Iran and England in a way that was not too harmful to England and would also secure a place for US oil companies (without endangering US oil interests in other countries around the world). This did not stop the United States from providing $145 million between 1953 and 1957 to shore up the Shah's regime. In all negotiations, England was insistent that no contract should be signed with Iran which gave a larger share to Iran than the existing contract between England and Iran. On December 6, 1953 a consortium of the largest oil companies of the world at the time known as the "seven sisters" signed an agreement in the presence of the leaders of the United States, England, and France (Fakhreddin 2003, Ferrier 2010, Galpern 2002, Gavin 1999, Kazemi 1985, Marsh 1998, Rahnema 1990). Iran was to be notified of the terms of the agreement. Soon afterward, vice president Richard Nixon and his wife arrived in Tehran and brought a twenty-member team consisting of economic, military, and oil experts. Appropriate deals were made to provide military assistance and equipment to Iran to prop it up in front of the Soviet Union, and also to secure the oil deal.

The negotiations with consortium delegates began on April 4, 1954 in Tehran. The first round was not successful because the delegates did not accept Iranian terms. The second round which took place in London was successful and the oil and gas agreement was signed on September 19, 1954. The lion's share (40%) of the newly formed consortium belonged to the Anglo-Iranian Oil Company. The second largest share holder was Royal Dutch Shell (14%). Five major and one small US companies accounted for 40% of the shares and the remaining 6% went to France (Fakhreddin 2003, Ferrier 2010, Kazemi 1985, Galpern 2002, Gavin 1999, Marsh 1998, Rahnema 1990).

The allies wanted to keep Iran within their influence after the end of the war to deprive the Soviet Union of a potential client-state like those in Eastern Europe, as well as the oil wealth and access to the Persian Gulf ports. Note that England, which had sole control over all southern-region oil in the past, was forced to settle for 40%, while the United States, that had nothing prior to negotiations, managed to carve up 40% of the share. In addition to all the oil and money that accrued to the members of the Consortium, Iran directly had to pay £76 million to the Anglo-Persian Oil Company, plus £670 million over 10 years. Indirectly, and through the Consortium, Iran ended up paying an additional £240 million. All of these were paid to "get back" Iran's own oil to her so it can sell it to the Consortium (Fakhreddin 2003, Ferrier 2010, Kazemi 1985, Galpern 2002, Gavin 1999, Marsh 1998, Rahnema 1990).

Although Iran was a "member of the Allied Forces" that won the war, in reality it never reaped any benefits from the allied victory. This Consortium, which was portrayed as a victory for Iran (about which the Shah used to boast about bringing multinational oil companies to their knees (Pahlavi 1980) was no different than the concession the Qajar shahs made to Britain and Russia. The new contract was much more beneficial to Iran than the old concessions of the Qajar Dynasty; however, this was not due to the Shah of Iran's ability to get a better deal. This contract did not give as much share to Iran as other countries were receiving for their oil. Article 41 of the Consortium agreement states that, "The government of Iran cannot change any provision of the agreement or cancel the agreement through any legislative or administrative action." In return, the Shah was given the crown and job security. According to Ramazani (1975), Iran and the United States agreed to keep parts of the contract secret. Since provisions like the English version would be considered the final or official version would have been very unpopular with the Iranian people (Article 48). The advantage of the Consortium contract for Iran was the fact that England no longer had power over Iran's oil. The contract solidified the role and influence of the United States in Iran.

Iran became a major (at least regionally) player in the containment strategy suggested by George Kennan (Abrahamian 2008, Daniel 2001, Gause 1985, Pollack 2005). Later, this strategy became the foundation of the United States foreign policy of President Truman. Kennan's views were instrumental in shaping the Cold War, and he was one of the main proponents behind the Marshal Plan to revitalize Europe. In pursuit of this policy and to fill the voids created by the removal of England from many of its former colonial regions, the United States (in 1952) proposed the Middle Eastern Defense Organization (MEDO). The main purpose of the organization was to keep Egypt from falling into the Soviet Bloc, assure the security of the Suez Canal, and to provide secure passage for Middle East oil. Turkey sponsored MEDO, but even Egypt (which was supposed to be a founding member) did not sign it. Gamal Abdel Nasser had leftist ideas and became the second president of Egypt in 1956. He was against this idea, which he considered a tool to further the imperialist reach of the United States. In 1953, with the end of Truman's term (and due to the lack of support from most Arab countries), MEDO became meaningless.

Events in Neighboring Countries

In the rest of the region, Egypt was very unhappy with the presence of England on its soil. So, Egypt refused to be party to any treaty, defensive or otherwise, as long as England was part of it. On the other hand, Turkey was still leery of the

Soviet Union and could not forget the Soviet's attempts to control Bosporus and Dardanelle at the end of the Second World War. Having lost its imperial power Turkey was very eager to enter into a defensive treaty. In January of 1955, the president of Turkey (in a trip to Washington D.C.) convinced President Eisenhower to agree to a defensive treaty between Turkey and Pakistan, which materialized in April. The Soviet Union and India both expressed great opposition, each for their own reasons. Within a few months, Iraq, England, Turkey, Pakistan, and Iran were all members of this new defense agreement. The defense agreement was a weak copy of North Atlantic Treaty Organization (NATO), but it was the weakest of such treaties designed to curb the Soviet Union's expansion. Some argue that England wanted to be part of the treaty, which by now had changed its name to the Central Treaty Organization (CENTO), in order to intervene in the affairs of the region. Britain's influence had been waning in the region for some time. The United States, however, did not join because it was worried that (in case of an Arab-Israel war) it would be obligated to intervene on behalf of the Arabs. Nevertheless, this treaty completed the ring around the Soviet Bloc in accordance to Kennan's containment doctrine.

In 1955, England, which had many military bases in Iraq, signed an agreement of collective security. On July 26, 1955, President Gamal Abdol Naser of Egypt nationalized the Suez Canal (Abrahamian 2008, Daniel 2001, Galpern 2002, Gause 1985, Marsh 1998, Pollack 2005). On October 21, British and Israeli leaders met to "solve the problem." Of Nasser and his threats to nationalize the Suez Canal. On October 29 Israel attacked Egypt. Two days later British and French bombers destroyed the Egyptian airfields. Five days later, they landed troops in Port Said. Soon, Israel won the war. Inspired by the Egyptian experience some Iraqi army officers revolted and overthrew the Iraqi government on 14 July 1958. As a result of the revolution Iraq left CENTO, weakening it even more. The leftist tendencies of the new Iraqi government increased the importance of Turkey, Iran, and Pakistan in blocking the Soviet expansion.

The relationship between Iran and the United States became stronger as time went by. Numerous treaties and agreements were signed. These treaties, which started as early as 1949 with a cultural exchange agreement, included all aspects of international relations between the two countries. Other treaties involved a loan, agriculture, aviation, energy, and (in 1957) an agreement was signed to cooperate in development of non-military nuclear power (Abrahamian 2008, Daniel 2001, Kibroglu 2006, Pollack 2005). On December 14, 1959, Iran and the United States signed a joint defensive agreement guaranteeing the United States' commitment to Iran's security ("Chronology of Iranian History Part 3", United States 1959, 1960). The United States' advisors were everywhere and were shaping the Iranian government based on a US model.

In 1957, with the help from the CIA, FBI, and Mossad, Iran created a National Security and Information Organization (Sazman Etelaat va Amniyat Keshvar or SAVAK), which became one of the most notorious of such organizations around the world, and at times, one of the more effective. SAVAK's first task was to destroy the remnants of the Tudeh Party using the 1931 decree banning socialist ideology. Once SAVAK proved effective, its objectives were expanded to attack the National Front (Gasiorowski 1991, Haqsenas 2010, Pollack 2005, Sheikoleslami 2010, Yarshater 2006, Zabih 1988). Later, SAVAK was charged with handling other so-called terrorist groups that emerged. Whatever little compassion that was left for the United States after the *Coup d'état* against the Iranian people and their beloved Mosaddeq was lost with SAVAK's establishment. Its relentless arrests, tortures, and murders of Iranian citizens were taken as intolerable. The sentiment was also extended towards Israel as well. Since Iran and Israel do not have borders and neither is a major global power, there should have been no animosity between them, except perhaps, indirectly through the cry of Palestinians and other Arab countries.

The "loss" of Iraq, despite all of the treaties with the United States, England, and regional powers (CENTO), sounded an alarm and rethinking of the US foreign policy. The change was made more plausible after the election of President Kennedy, who was in favor of a reform-based policy to curb the spread of communism. The analysts had concluded that the main source of the spread of communism was the lack of democracy and denial of basic living standards for the masses in third world countries. The prescription was to reduce dictatorship; increase distribution of wealth and income; respect human rights; permit more open media; and reform economic structures. President Kennedy was hoping to cut military aid to United States puppet regimes and replace it with economic assistance (Abrahamian 2008, Daniel 2001, Pollack 2005). In order to limit the spread of communism the United States had propped up numerous puppet, corrupt, unpopular, and dictatorial governments around the world. This was mostly accomplished through *coup d'état* and military force followed by massive transfer of military hardware and advisors. Consequently, the United States began pressuring the governments of these countries, including the Shah, to change and begin conforming to the new doctrine. Again, it was not the interest of the people of these countries that guided the democratization and economic development programs; rather, it was the ineffectiveness of the previous American Foreign policy to limit the spread of socialism that dictated the change. Here again the United States was ordering its puppet regimes to do what was in the interest of the United States and consequently was ill-received by the masses of these countries.

During the 1950s global oil production outpaced global demand and the price of oil was falling both in real as well as nominal terms. Substantial decline in oil revenue for Iran, during and after the nationalization of oil (followed by price declines) caused major financial problems. These problems were but exacerbated by the importation of luxury products and consumer goods that competed with and destroyed domestic production. Substantial transfers of currency to abroad further fueled this crisis. Furthermore, the seven-year development plans proved very expensive and the military budget was growing at a rapid pace. Both of these expenditures were highly inflationary, the results of which became evident in 1957 when prices began a rapid spiral upward. Between 1957 and 1960, prices increased by some 35 percent (Abrahamian 2008, Amuzegar 1992, Central Bank of Iran 1970, Daniel 2001, Karshenas 1999, Nowshirvani 2010, Pollack 2005, Razi 1987, Rahnema 1990).

Due to the financial crisis Iran requested a loan from the International Monetary Fund (IMF). IMF was established by the United States in 1944 to provide a source for funding different projects and activities around the world. This was a different source of funds than in years past. Before, major countries provided loans to smaller countries, and mostly to their puppet regimes. The people of those countries realized the harmful effects of such loans and were against it. The new lending agency did not have such stigma attached to it. However, careful examination of the rules and regulation and sources of funds clearly demonstrates the influence of the United States on the IMF. With a positive nod from the United States, the IMF agreed to give a $35 million loan, but as had become customary, it placed some terms and conditions upon Iran. Specifically, it required a salary and wage freeze, and a budget trim. The United States offered to increase the amount to $85 million if Iran agreed to proceed with the land reform, which had already been through several stages and had the support of the liberals in the cabinet. Public dissatisfaction with the situation resulted in an increase in demonstrations, which in turn, convinced US advisors that it was necessary to have some sort of reform in the country.

Pressure to Reform

The Shah, under pressure from the United States, allowed independent candidates as well as the candidates from the National Front to participate in the election for the 20th Majlis in 1960. Egbal was replaced by Sharif Emami as the Prime Minister. Sharif Emami enforced a belt tightening measure.

In 1960, the Shah proposed land reform in the form of selling land to the peasants that were cultivating the land. The Shah sold some of "his lands" to the peasants and was hoping that other major landowners would follow suit. At the time of the first sale of land by the Shah, Mosaddeq and the National Front were of the opinion that those lands were not the Shah's to sell in the first place. Previously, Mosaddeq had argued that Reza Shah took that land by force and that it had to be returned to their respective rightful owners.

A bill was drafted at the Majlis for land reform. The bill faced two problems. One problem was that the majority of the members of Majlis were large landowners who began introducing exceptions and modifications to the point that the finished product, if passed, would not have had any major impact. The second opposition came from Ayatollah Borujerdi who was the undisputed religious leader of the country at the time and was residing in Qom. The land reform bill passed by the Majlis in May 1960, and immediately was denounced by Ayatollah Borujerdi. Consequently it accomplished nothing.

Nine months after Sharif Emamy took office a major confrontation of government employees, especially teachers, with the police took place in front of the Majlis (Abrhamian 2008, Daniel 2001, Pollack 2005). Subsequently, Sharif Emamy was replaced by Dr. Ali Amini. Dr. Amini had a good relationship with the United States (he had been Iran's ambassador to the US) and he had a good relationship with the National Front (he was Mosaddeq's finance minister). He was a nationalist and reform minded (he had been advocating land reforms since the end of the WWII). However, the Shah disliked and distrusted Dr. Amini, maybe for the same reasons. Dr. Amini lasted a little over a year, because he failed to get US support, and clashed with the Shah over military budget cuts. He was disliked by the public because he advocated the implementation of the IMF's austerity measures, and he failed to get the support of the National Front (because he did not dismantle SAVAK and did not hold free elections).

Apparently the Shah was not clear about the new direction of foreign policy of the United States, because he sent an envoy to the United States in 1961 to negotiate more weapons sales to Iran. The envoy was received poorly and was unable to meet with the president for three weeks (Gause 1985, Looney 1988, Pryor 1978, Ronfeldt 1978). In 1961, Ayatollah Broujerdi passed away and room for other clergy, especially Khomeini opened up. A United States envoy came to Iran in February 1962 (right after Dr. Amini became the prime minister) and encouraged the Shah to focus on peasant issues, land reform (which became very popular around the world in the 1960s), economic planning, and a more open political climate. Ignoring the recommendations, the Shah travelled to the United States in April to secure military and economic aid.

In August, Vice President Johnson travelled to Iran and finally managed to force the Shah to give in and agree to pursue the new approach. The reforms were packaged as modernizing the country and wrapped in claims of openness, democratization, and economic justice for the masses, mimicking communism slogans and ideals while trying to take away communist appeal. In return, the United States promised to support Shah no matter what (Abrahamioan 2008, Daniel 2001, Pollack 2005). Meanwhile, election laws were modified in the fall of 1962 to exclude the requirement that elected officials be sworn into the office upon the Qoran, as required by the constitution. The motive for doing so is not clear. It offended many orthodox believers and a majority of the liberals. Needless to say, this was opposed by Ayatollah Khomeini who was gaining popularity since the death of Ayatollah Boroujerdi.

After Ayatollah Boroujerdi's death the Shah, who had a showy relationship with Khomeini, sought to legitimize his regime in the eyes of the public (especially the faithful). The Shah did not acknowledge any religious leader and may have been hoping that their importance would vanish. However Ayatollah Khomeini questioned the claimed spiritual link between the Shah and Imam Reza (Enghelab 1963). Thus, the regime was denied any religious foundation. Ayatollah Khomeini gained popularity due, in part, to his relentless attack on the Pahlavi Dynasty. Among the believers, especially his followers, he was the only one at the time that relentlessly opposed the Shah. This coincided with the decline of the Tudeh Party, mostly due to numerous arrests and exile, both voluntary and forced. The National Front was in disarray due to Mosaddeq's demise. Other groups came into the picture yet were not important at this time.

Iran's "White Revolution" began by land reform, sales of stocks of government-owned factories, creation of a literacy corps, and freedom of the press (Pollack 2005). The US Secretary of States' office, in a memorandum to the White House dated January 21, 1963, stated that things were improving in Iran and the Shah's chance of survival had increased. On January 22 Khomeini, after securing the support of the other religious leaders of Qom, denounced the Shah's dependence on foreigners and support from Israel. Khomeini ordered a boycott of the referendum on the White Revolution (Pollack 2005). Many years later, this denouncement seemed to have helped shape the foreign policy of the Islamic Republic with the United States and Israel.

On January 24, an army column was moved to Qom and the Shah delivered a harsh speech against religious fanatics. On January 26, the White Revolution was put on a referendum and was approved by the majority. Under the provisions of the White Revolution, land was purchased from landlords and was sold to peasants at 30% of market value on 25 year terms. Some land owners, especially the clergy, were not happy with the deal. In the end, 9 million peasants (or 40% of the population of Iran in 1963, but a higher percentage of the population if family members were counted) became landowners. Other aspects of the reform included privatization of government owned enterprises; profit sharing for workers; voting and other rights for women; formation of a literacy corps; a health corps; and a reconstruction and development corps (Abrahamian 2008; Pollack 2005). These and other articles added later were progressive, even in today's standards. Many of them mimicked institutions established by the United States, such as the Peace Corps, except they were formed by the domestic government, operated within the country and maintained legal authority. This was not the case for the Peace Corps. The Peace Corps is composed of volunteers, who were not given power or authority, working in other countries. The White Revolution promised to improve the lives of the average Iranian, but implementation became an issue. It also increased opportunities and expectations. Some argue that the latter was increased more than the former, and thus created discontent among Iranians.

On June 3, Ayatollah Khomeini delivered a fiery speech at Feyziyeh Madreseh (a religious school in Qom) and attacked Reza Shah, the founder of the Pahlavi Dynasty. In the middle of the night of June 5 Ayatollah Khomeini was arrested. This led to massive demonstrations in Qom, Tehran, Varamin, Mashhad, and Shiraz in the following days (Abrahamian 2008, Daniel 2001, Pollack 2005). According to some accounts, more than 15,000 people were killed in clashes between demonstrators and the armed forces. It took six days for the government to control the situation. The irony, at least in the eyes of the Iranians, was that President Kennedy's administration was advocating human rights, democracy, and political freedom; but in Iran, staying silent about the Shah's repressive policies.

The show of force by the Shah's military and US weapons was temporarily effective. It resulted in keeping the Shah in power at the time. The proclamation of Ayatollah Khomeini, and the mass uprising against the regime, was the first step toward a movement that would cause the collapse of the Shah's regime some 25 years later. Soon, the Shah proceeded with numerous acts of change which at the surface looked progressive, but in reality were designed to consolidate his power and make sure that the relationship with the United States remained mutually beneficial (Abrahamian 2008, Daniel 2001, Hart 1980, Kurzman 1995, 1996, Nikazmerad 1980, Pollack 2005).

Escalation of Problems

It is not clear why the United States began demanding the establishment of capitulation in Iran. It could be that the unrests of June 5, 1963 played a role, or like other major powers before, it could not resist demanding special privileges. Perhaps the government of the United States was concerned about the public opinion of American voters in case of trial of American citizens in Iran. The United States demanded that its advisors, military personnel, civilian workers, and their families be exempt from Iranian laws. This, in effect, demanded the establishment of a capitulation law which was popular with colonial and imperialist powers of the past. The Majlis had banned capitulation in May 1927.

In the middle of all this, the relationship between Iran and the Soviet Union improved, especially when President Brezhnev visited Iran on November 16, 1963. This might have been interpreted as a balancing act between the superpowers. One should not forget that at the same time the Shah's regime was actively combating the communists in Iran, especially the Tudeh party, and was becoming closer and closer to the United States.

In December 1963 the Shah ordered the creation of a new political party to be called the New Iran Party (Hezb Iran Novin) to replace the National Party. The public considered this a US plot for better control of the Majlis. The party head was Ali Mansur, whose family had long and close ties to the Shah's court and was active in Iranian politics. The two political parties were both government operated; however, the creation of the New Iran party was the first step to revert to a one-party system for tighter control. Although the Shah's regime controlled the elections to make sure the "right people" were in the Majlis (and the Shah actually appointed half of the Senators, mostly retired high ranking military); nevertheless, it could not afford to take a chance with passing the capitulation law. Although the US State Department wanted immunity from prosecution to be signed informally between the Foreign Ministry of Iran and the U.S. State Department (Foreign Office, 1964a), Prime Minister Mansur believed this to be a constitutional issue which had to go through the Majlis. Therefore, on October 3, 1964 the Majlis had a long day, passing the submitted capitulation law at midnight. Of the 200 members in the Majlis, only 130 were present. Sixty members of the Majlis openly opposed the law. Eventually the law passed but its extent was limited to US workers only, not their families (Abrahamian 2008, Daniel 2001, Pollack 2005). Although every effort was made to keep this a secret, one of the workers in the Majlis leaked the bill to Ayatollah Khomeini who was released from jail in April. Soon the Majlis approved a $200 million loan from the United States (Foreign Office 1964a). In a speech on November 5, 1964 Ayatollah Khomeini revealed the secret agreement and denounced the Shah and the United States's conduct in Iran. Ayatollah Khomeini proclaimed that the $200 million "loan" was the reward that the Shah received for selling Iran's sovereignty and independence. In his speech he clarified and pronounced the meaning of the recently passed law:

> If some American's servant, some American's cook, assassinates your Marja in the middle of the bazaar, or runs over him, the Iranian police do not have the right to apprehend him! Iranian courts do not have the right to judge him! The dossier must be sent to America, so that our masters there can decide what is to be done! .. They have reduced the Iranian people to a level lower than that of an American dog. If someone runs over a dog belonging to an American, he will be prosecuted [Algar 198: 181-188].

This law evaporated any remaining love or respect for the US among the majority of Iranians. The exile of Ayatollah Khomeini paved the road to his acceptability by a larger number of Iranians.

In January 1965 Ali Mansur, the Prime Minister, was assassinated by a member of an Islamist student organization. The assassination of Mansur, instead of being a warning, made the Shah even more determined to consolidate his power and get closer to the United States (Abrahamian 1970, 1980, 2008, Daniel 2001). Having failed to gain any popular basis among the people he could only count on US military muscle to stay in power. He appointed Hoveida as Prime Minister. Hoveida was an unknown and a small player in Iranian politics, practically a "nobody" in politics. The Shah chose him as the prime minister to demonstrate that no other person in government mattered. He was to rule with an iron fist, even if the most inept person was to be the "head of the government." The Shah kept Hoveida in his position for 13 years until the first breezes of the Islamic Revolution forced changes in the status quo.

It seems that at least the Shah was convinced of his appointment by divine forces to rescue Iran from whatever perceived inadequacies that it might have had. More and more the Shah was convinced that he was a great man and that the nation adored him (Abrahamian 2008, Pollack 2005). For example, on the 25th anniversary of his rule in 1965 he forced the Majlis to give him the new title of "Aryamehr" which means beloved by Aryans, the primary ethnicity of Iranians. Soon, the Iranians twisted the title to "Ary as mehr" which means the one who lacks affection. Interestingly, in the same year, one of the members of the Royal Guard, which had sworn to die defending the Shah, attempted to kill him (Abrahamian 2008, Ansari 2003, Daniel 2001, Pollack 2005). Furthermore, the public did not show any interest in participating in any celebration. Many people, especially the elite and students, were actually critical, while the general public could care less. The only participants in the official ceremonies were government workers who had to leave their jobs to partake in the "ceremonies," which agitated their patrons when their jobs were left undone. The store owners were asked to show their patriotism by putting up decorations, especially colored lights, and stories about SAVAK's forceful tactics added to the public's discontent.

As time went by the Shah's arrogance increased and he moved more towards dictatorship, much in the same way as his father. Rumors of his affairs with young women were widespread. Other rumors claimed his isolation from people and decision making without consulting anyone, except possibly a small group of insiders. In 1966, the Shah wrote his second book named *The White Revolution*, which became mandatory reading for all eighth grade high school students in a course bearing the same name. The Shah's ego was exacerbated by constant and unequivocal approval of whatever he said or did by his cronies and servants, as he called his officials. A good example of this is evident in several speeches by Prime Minister Hovieda who stated "the secret of Iran's economic and social success lay in the fact that it did not follow baseless schools of thought, nor was it inspired by East or West in its revolution - the revolution was inspired by national traditions and the Shah's revolutionary ideas" (Ansari 2003). We will deal with Shahs economic success in chapter five.

A notable point here is that the regime realized that it is not possible to conceal the inherent conflicts of revolutionary monarchy without creating a new and different ideology. Some books were written about this "unique ideology" which we will not mention here to avoid giving any credence to such worthless trash. More and more Shah was considering himself a gift to the Iranians and acting as if the country would not survive without him, much in the same way as his father did before him. The notion of neither East nor West became a major slogan later on when the Islamic Republic came to power. This shows how much the Iranians would like to determine their own destiny rather than being a follower of others. Both Mohammad Reza Shah and Ayatollah Rohollah Khomeini understood this and utilized it, with different levels of sincerity. This reality has been either not understood by the United States or has been thought to be insignificant in US-Iran relations.

The Calm before the Storm

The decade of the 1960's was devoted to consolidation of the relationship with the United States and centralization of power in Iran. The Shah managed to destroy the National Front party through *coup d'état* and the pursuant cleansing which included jailing prominent leaders. The Tudeh party had been disabled by more effectively carrying out the 1929 law, which made the Tudeh Party illegal (Ghods 1990). By this time the majority of the Tudeh party's leadership was dead, in prison, or in exile (Abrahamian 1978, 2008, Haqsenas 2010). The political climate had been neutralized by creating the Iran Novin Party which had effectively (and later on literally) converted the country into a single political party, which was under the full control of the Shah. Not only the Shah but many within, as well as without the country, had been convinced that the Shah (and by extension the Pahlavi Dynasty) was safe and here to stay for the foreseeable future.

During the decade from 1963-1973 Iran's economy grew at an average rate of 10% (Ansari 2003). Supporters and admirers attributed the rate to the Shah's revolution and wisdom. Opponents pointed out that the growth was due to increased oil income which was distributed very unevenly. In this period, being rich in Iran was a unique experience that was not comprehensible even by rich people in most other places. The limited rich in Iran, which by most accounts were limited to fewer than 1,000 families, meant access to things that were not even dreamed or imaginable by most other Iranians.

Many grand and exorbitant projects were undertaken such as great dams and power companies. Iran even began building nuclear power plants in order to be ready when its oil ran out (Kibroglu 2006). Oil revenues increased even more after the Yom Kippur War and the Arab Oil Embargo. Interestingly, the Shah claimed to have been the main reason for the price increase brought about through OPEC after the oil embargo (Pollack 2005). The regime continued to showcase former enemies, especially from the communist Tudeh party, who were "confessing" to their misled life of the past and praising the regime for surpassing what they were dreaming to accomplish through communism. Many of those that confessed and joined the chorus of praise obtained government jobs. In fact many intellectuals were persuaded to keep quiet by offering them lucrative government positions. Those who did not participated were dealt with through SAVAK, which had become even more powerful.

The country was prosperous, the Shah's enemies were silenced, and the future looked bright for the regime. Therefore, in 1967 the Shah decided to coronate himself and his wife. He expressed that he wanted to delay his coronation until the country had been stabilized. The ceremony, allegedly a unique Iranian tradition celebrating the kingdom, turned out to be a copy of British appropriation (Wright 1996). The ceremonies were enjoyed by the middle and upper classes, the former regarding it as a soap opera while the latter saw an opportunity to demonstrate to the world Iranian modernization as well as a good party. The poorest and lowest income people, however, considered this yet another Western extravagance that they wished to have nothing to do with. Some were disgruntled by the expenditures, which they considered their share of national wealth (Hart 1980, Pollack 2005, Razi 1987). Another new twist in the royal succession was the coronation of the queen, who through amendments to the constitution, was to succeed the Shah in case of his death should their son still be a minor. The Shah presented this as evidence of emancipation of women by him, ignoring the fact that numerous queens had ruled Iran in its long history.

Once self pride set in, and the Shah was able to thank himself for what he had accomplished for the nation without anyone objecting or even questioning the notion, then it seemed that for him the sky was the limit. It could be that he really believed that the citizens could not get enough of him, so he began planting statues of himself around the country. Within three months six statues of the Shah were erected in squares in different cities. One can argue that the country was prosperous and could afford a little indulgence; or that these cities did not have any statues and that it was the peoples' right to have a statue of Shah in their main square. The only problem is that the same statue-creating behavior is demonstrated by other dictators around the world. The idea, that people demanded the statues was dismissed, when no sooner the Shah left the country's air space the same people pulled his statues down. Such has happened to many other dictators around the world before and after the Shah.

The more upset the people were becoming with injustice and dictatorship the greater the statements of grandeur by the Shah's servants were becoming. For example, the Premier Hovieda stated that the "entire world is studying and imitating what has been achieved through the Shah's genius." and "The Shah's era is the most brilliant era in Iran's history (Mardom 1967)." It seems these people had forgotten how wide-spread the agrarian land reform was in the 1960's or the fact that all such activities were attempts to curb the spread of communism with (not so subtle) recommendations from the United States. By 1970, the signs of inflation and slower growth were emerging, which seemed not to alarm anyone. The upper and, to a lesser extent, middle classes were riding high on the wave of expansion. A minor slowdown in growth or a little inflation was of no concern.

Regional Might and Supremacy

Around this time there were rumors about "granting" independence to Bahrain. The majority of Iranians did not appreciate the talks of separating Bahrain from Iran. Bahrain, which historically was part of Iran, had been occupied by the colonial forces of Portugal, Spain, and Britain. Britain, depending on its needs or abilities at the time, returned Bahrain to Iran, occupied it, or allowed "independent" Sheikhs to rule the island (Abrahamian 2008, Daniel 2001, Khadduri 1951, Pollack 2005). In 1968 England announced that it would no longer keep a military presence east of the Suez Canal. The United States, which was filling the void left by British military withdrawals around the world, was ready to step forward (Abrahamian 2008, Daniel 2001, Meskill 1995, Pollack 2005). The United States suggested an independent country be created on the island. Although the majority of Iranians were against it the Shah succumbed to US pressure and agreed to allow the United Nations to conduct a referendum in Bahrain. Although Bahrain was listed as the 14th province of Iran in all the geography textbooks of the country it had been operating "independently" under the British protective umbrella. Bahrainian people voted to become independent and United States money and advisers poured into the tiny island and it became one of the major US naval bases in the region. Later it played a major role in US military operations in Kuwait in 1991 and later in the attack against Iraq in 2003.

If Bahrain had remained part of Iran, then by international maritime laws the entire Persian Gulf, with the exception of a short and un-navigable strait south of the island, would have been considered Iranian territorial water (Abrahamian 2008, Daniel 2001, Khadduri 1951, Meskill 1995, Pollack 2005). This would have deprived Iraq and Kuwait from access to international waterways without going through Iranian waterways, hence the need to have Iranian permission. It is worth noting that most of Bahrain is south of the tip of Qatar. The United States could not risk that much oil flow in control of a single country, even if it was the best ally in the region. The fear of having an unfriendly government in Iran came to realization in 1979. The loss of Bahrain laid heavy in the hearts and souls of Iranians, who for the past 200 years had seen parts of the country separated or out-right handed over to foreign masters of different Shahs. Consequently, the opposition to this particular Shah and his master, the United States, deepened.

Whatever the issues, apparently they were not significant for the regime, which decided to celebrate 2,500 years of Iranian monarchy. The idea that had surfaced in the 1950s was to highlight Iranian culture and its contribution to the world. It was planned to be a boost to Iranians and give them a sense of pride that they deserved. Instead the plans turned out to be focused on the Shah, boasting him as the same as Cyrus the Great, and to showcase newly built structures such as the Shahyad (meaning Memorial of Shah), structure which was located in a huge square with the same name (now the square is named Azady, or freedom). The building and the square were finished in time to greet incoming dignitaries who landed at the nearby airport. Other more useful structures such as dams were also to be featured.

In reality, however, the ceremonies became a party for kings, presidents, and other dignitaries of countries to indulge on imported delicacies (of which most Iranians were unfamiliar). With few exceptions from the highest ranking members of the 1,000 families the rest of Iranians were not allowed anywhere near the ceremonies. While there were numerous objections from disgruntled merchants, office workers, and the general public, the main opposition came from Ayatollah Khomeini who used the occasion to declare his opposition to the monarchy for the first time. He no longer was willing to settle for liberty or human rights under the constitutional monarchy. In 1971, for the first time since before the constitutional revolution of 1905, a prominent member of the clergy was no longer accepting the presence of the monarchy, in any shape or form. It seems absurd for a ruler to lose a part of his country and celebrate his greatness months afterward.

None of the western media published Ayatollah Khomeini's opposition, while almost all praised the Shah and his accomplishments; without being aware of the true sentiment of Iranians, *The Times* (1971) stated, "To the people of Iran, the Institution of Monarchy is not a mode of government but is rather a way of life which has become an essential part of the nation's very existence (Abrahamian

2008, Daniel 2001)." Within seven years the world realized how far this assessment of Iran and Iranians was from reality. During 1971 the idea of the Great Civilization the continuation of the White Revolution was also introduced by Shah. The Shah boldly declared that within 12 years, Iran would achieve the Great Civilization, which would signify the kindest welfare state in the world without a single illiterate person. The atmosphere in Iran after the loss of Bahrain and the extravagant ceremonies of the 2,500 years of monarchy rule was very negative. Small pockets of opposition in the form of guerilla warfare by young Communist as well as Muslim groups were emerging (Abrahamian 1978, Pliskin 1980, Pollack 2005).

Suddenly, in 1971, it was announced that Iran had reclaimed three of its islands in the Persian Gulf that were occupied by the United Arab Emirates (UAE). They were called Abu Musa, the Greater Tunb, and the Lesser Tunb (Abrahamian 2008, Daniel 2001, Ramazani 2010). The British had an agreement with the Emirate Sharjah about the islands, which were becoming part of the United Arab Emirates. But Iran declared that unless the three islands remained with Iran, it would refuse to recognize the formation of the UAE. Finally, England, which was anxious to get out of colonial territories, and the UAE agreed to give up the disputed islands. In reality, it was the United States that gave the final nod, which may be considered as a consolation prize for separating Bahrain. The move could have been due to an agreement to have Iran fill the void created by British withdrawal from the region. In light of what was going on in Vietnam, it was better for the United States to allow someone else to play the role of the gendarme of the region. The return of the three tiny islands to the mother land was broadcasted in earnest. Iranians celebrated the occasion, not for the possession of the three little islands but rather in the faint hope that the era of decline had come to an end. The regime and its media glorified the occasion to its fullest extent, not realizing that it was another nail in the coffin of the Pahlavi Dynasty.

The Shah's opponents used the occasion to point out that the Shah was simply subservient to foreign interests and was doing the dirty jobs of foreigners using Iranian money. The fact that Iranian troops were fighting against Marxist rebels (supported by South Yemen) in Dhofar, Northern Yemen in the later 1960s contributed to the above feeling as well as convincing President Nixon of the United States that the Shah could fill the void left by British withdrawal.

Even worse was Iran's engagement in Oman, which started in 1970 and lasted until 1975 (Gause 1985, Looney 1988, Pollack 2005). While Iranian troops and equipment were involved the engagement was not officially announced. Over time more and more people in Iran became aware of the involvement in Oman. Iranians were asking why the blood of their youth should be shed fighting for United States interests in the region. Gradually, numerous anti-government guerilla warfare groups were formed. The most successful ones were Chirik-haye Fadaee Khalq Iran (a Marxist group also known as Fadayianeh Khalqh) and Sazemaneh Mojahhedineh Khalqh Iran (an Islamist-Marxist group), both of which were popular among university students. In February 1971, a group of guerillas attacked the gendarme station in the village of Siahkal in the forests of Gillan and killed three gendarmes. This resulted in the escalation of arrests of the opponents. Many of the arrested people were later put on display to allow them to renounce and denounce their behavior and ask for clemency and forgiveness. The number of people who were put on display was a very small fraction of those arrested. One of those on display, Khosrow Golsorkhi, was immortalized by the fact, when he was put on display in 1974 he refused to recant and attacked the Shah's regime and what it was doing to Iran and Iranians. Needless to say the charade ended and no more "confessionals" were showcased (Abrahamian 1980, 2008, Daniel 2001, Pliskin 1980, Pollack 2005).

During this period the relationship between United States and Iran was the coziest. The Shah and President Nixon were very close. The United States was having difficulty in Vietnam and could not directly get engaged in other hotspots; Iran was receiving much oil revenue; the Shah was ambitious to become the regional superpower; and there were many Marxist uprisings around the Middle East. Therefore, Nixon agreed to sell any and as many weapons systems that Shah wished to purchase, except for nuclear weapons (Pollack 2005). The Arab Oil Embargo of 1973 and the pursuing strengthening of OPEC provided the means to heavily arm Iran. After the sudden jump of oil prices, the Shah was portrayed as the driving force who managed to force the Western powers to cave under his pressure and agree with price increases. On the other hand, in the international arena, he portrayed himself as the best ally of the West, since he increased Iranian oil supplies while the Arab Oil Embargo was in effect. The fact that these two claims cancel each other, and that it is not possible to increase the price of any commodity by increasing its supply, did not seem to bother the official propaganda machine. Most of the increased revenue was used to purchase US weapons systems used to fight the insurgency in Oman, which was not a national security concern of Iran. Another sizable portion was used to purchase bankrupt and failing industries in the West, such as (a substantial part of) the Krupp steel factories in Germany.

In 1975 the Shah replaced the Iran Novin Party with the Rastakhiz (resurrection) party and officially created a single-party country. Once again, the crafty Iranians twisted the name a little to make it mean the "party of national excuse." The main difference between the Iran Novin and the Rastakhiz parties was that the latter required active participation from all the citizens. In the party's newspaper the Shah separated Iranians into two groups; those who believed in constitution, monarchy and the White Revolution and those who did not. The believers had to join the party. Non-believers were divided into two subgroups: the subgroup that belonged to illegal organizations (mostly he meant communists and socialists) and the subgroup that simply did not believe in the above principals. The former were told to take advantage of a free one-way ticket out of the country. The latter had to pledge allegiance to the country. Thus, for the first time, it was not sufficient to be a born citizen of the country; people had to join the Rastakhiz party, leave the country, or pledge allegiance. The above acts were conducted to give credence and legitimacy to the Shah's regime and force people to be active supporters of the regime, by membership in the Rastakhiz party, or alternatively by pledging allegiance to the Shah, to the regime, who believed he was the soul of the nation.

In 1976 the regime, in another attempt to solidify the prominence of the monarchy in Iran, decided to change the beginning date of the Iranian calendar from the pilgrimage of the Prophet Mohammad from Mecca to Medina to that of the inauguration of Cyrus the Great (Abrahamian 2008, Daniel 2001, Pollack 2005). The former, definitely has a foreign origin and the latter is absolutely pure Iranian. The problem was that the existing date had a religious significance for the people, while the latter created resentment from not only the faithful but also from intellectuals (for its historic inaccuracies and historical revisionism). In addition, the opponents of the regime also objected given they disliked whatever the regime did. Based on the new Imperial Calendar, the Shah's reign began on the year 2,500. The implication was that Cyrus the Great shaped the first 2,500 years of Iran's history and Mohammad Reza Pahlavi was going to shape the next 2,500 years. In 1977 the Shah released his third book named "Towards the Great Civilization" which was the theme that was being promoted by the regime. The theme implied that all problems have been solved and now was the time to finalize reaching the Great Civilization (which was predicted to occur within 12 years from 1976). The regime was insisting that the most central aspect of Iran and Iranian people was monarchy and the person of the Shah, without whom the country would not exist. However, suddenly in 1978 Iranians poured into the streets and by 1979 the Pahlavi Dynasty ended. In response to the Iranian Revolution the Shah escaped from Iran, Ayatollah Khomeini returned to Iran, and the Islamic Republic of Iran was established.

Conclusions

The zenith of Iranian-US relations came in 1973 when, in defiance of OPEC, Iran supplied oil in spite of an embargo by Arab countries and other OPEC members. Revenues increased dramatically and funds were used to purchase more advanced arms from the United States. The United States thought they had an everlasting friend in Iran and the Shah believed that his reign would last until his son Reza could come to the power and begin his reign. Ultimately both the Shah and the United States were ignorant of the forces unleashed in 1953. In 1979 relations reached their nadir as revolutionary fervor swept through Tehran, culminating in the taking of the United States' embassy by radical students and the holding of US diplomats for 444 days.

The roots of the current low-intensity conflict go back to this era. The social sentiment in Iran following the 1953 coup was one where they believed that little could be done to topple what the vast majority of the population believed was a corrupt and increasingly illegitimate government. While reforms introduced by the Shah were welcomed by many they tended to fall short of their goals. An example can be drawn from the "White Revolution" and its emphasis on land reform. While there were great expectations for progress little occurred, which alienated the majority of Iranians and leading to revolution.

Bibliography

Abrahamian, E., 1970. Communism and Communalism in Iran: The Tudeh and the Firqah-I Dimukrat. International Journal of Middle East Studies, 1(4), 291-316.

Abrahamian, E., 1974. Oriental Despotism: The Case of Qajar Iran. *International Journal of Middle East*, 5(1), 3-31.

Abrahamian, E., 1978. Factionalism in Iran: political groups in the 14th Parliament (1944-46). Middle Eastern Studies, 14(1), 22-55.

Abrahamian, E., 1980. The Guerrilla Movement in Iran, 1963-1977. MERIP Reports, (86), 3-15.

Abrahamian, E., 2001. The 1953 Coup in Iran. *Science & Society*, 65(2), 182-215.

Abrahamian, E., 2008. A History of Modern Iran, NYC: Cambridge University Press.

Amuzegar, J., 1958. Point Four: Performance and Prospect. *Political Science Quarterly*, 73(4), 530-546.

Amuzegar, J., 1992. The Iranian Economy Before and After the Revolution. *Middle East Journal*, 46(3), 413-425.

Bill, J.A., 2006. The Cultural Underpinnings of Politics: Iran and the United States. *Mediterranean Quarterly*, 17(1), 23-33.

Bill, J.A., 1988. *The Eagle and the Lion: The Tragedy of American-Iranian Relations*, New Haven, CT.: YAle University PRess.

Bonakdarian, M., U.S.-Iranian Relations, 1911-1951. In *The United States and the Middle East: Diplomatic and Economic Relations in Perspective*. Yale Council on Middle East Studies, pp. 9-25.

Bostock, F., 1989. State Bank or Agent of Empire? The Imperial Bank of Persia's Loan Policy 1920-23. *Iran*, 27, 103-113.

Carmical, J., 1951. World Eyes Iran on Oil Seizure Bid. *The New York Times*.

Chaqueri, C., 1999. Did the Soviets Play a Role in Founding the Tudeh Party in Iran? *Cahiers du monde Russe*, 40(3), 497-528.

Committee on Foreign Relations: United States Senate, 1960. THE MUTUAL SECURITY ACT OF 1960. *Public Law*, Mu(1319).

Cottam, R., 1980. American policy and the Iranian crisis. *Iranian Studies*, 13(1), 279-305.

Cremer, J. & Salehi-isfahani, D., 2010. The of Rise Oil and Prices : Fall A Competitive View. Noûs, (15).

Dadkhah, K., 2001. The Iranian Economy during the Second World War: The Devaluation Controversy. *Middle Eastern Studies*, 37(2), 181-198.

Daniel, C., 1952. U .S. and Britain Confronted By Dilemma on Help to Iran. *The New York Times*.

Daniel, C., 1951. British Warn Iran of Serious Result if She Seizes Oil. *The New York Times*.

Daniel, E.R., 2001. *The History of Iran*, Westport, CT.: Greenwood Press.

Davis, S., 2006. "A Projected New Trusteeship"? American Internationalism, British Imperialism, and the Reconstruction of Iran, 1938-1947. *Diplomacy and Statecraft*, 17, 31-72.

Edwards, A.C., 1947. Persia Revisited. *International Affairs*, 23(1), 52-60.

Ellender, A.J., 1960. *A Report on United States Foreign Operations*, Washington, D.C.: GPO.

Ellender, A.J., 1958. A Review of United States Foreign Policy and Operations. Washington, D.C.: GPO.

1954. Ex-Foreign Chief of Iran Executed. *New York Times*.

Foran, J., 1989. The concept of dependent development as a key to the political economy of Qajar Iran (1800-1925). *Iranian Studies*, 22(2), 5-56.

Galpern, S.G., 2002. Britain, Middles East Oil, and the Struggle to Save Sterling, 1944-1971. *Middle East*.

Gause, F.G., 1985. British and American policies in the Persian Gulf, 1968-1973. *Review of International Studies*, 11(04), 247-273.

Gavin, F.J., 1999. Politics, Power, and U.S. Policy in Iran, 1950-1953. *Journal of Cold War Studies*, 1(1), 56-89.

Ghaneabassiri, K., 2002. U.S. Foreign Policy and Persia, 1856-1921. *Iranian Studies*, 35(1/3), 145-175.

Greaves, R.L., 1965. British Policy in Persia, 1892-1903--I. *Bulletin of the School of Oriental and African Studies*, 28(1), 34-60.

Greaves, R.L., 1965. British policy in Persia, 1892-1903-II. *Bulletin of the School of Oriental and African Studies*, 28(02), 284-307.

Greaves, R.L., 1968. Some Aspects of the Anglo-russian Convention and Its Working in Persia, 1907-14--I. *Bulletin of the School of Oriental and African Studies*, 31(1), 69-91.

Greaves, R.L., 1968. Some Aspects of the Anglo-Russian Convention and Its Working in Persia, 1907-14--II. *Bulletin of the School of Oriental and African Studies*, 31(2), 290-308.

Greaves, R.L., 1986. Sīstān in British Indian frontier policy. *Bulletin of the School of Oriental and African Studies*, 49(01), 90-102.

Greaves, R., 1991. Themes in British policy towards Persia in its relation to Indian frontier defence, 1798-1914. *Asian Affairs*, 22(1), 35-45.

Halliday, F., 2008. From South Persia Rifles To North Persia Guerrillas: Sir Percy Sykes, Mirza Kuchik Khan and the Legacies of Che Guevara. *Asian Affairs*, 39(2), 173-188.

Hangen, W., 1954. Iran And Oil Group Initial Agreement To Resume Output. *The New York Times*.

Hart, J., 1980. A bibliographical survey of the Iranian revolution. *Iranian Studies*, 13(1), 369-390.

Hess, G.R., 1974. theiraniancriris194546.pdfThe Iranian Crisis of 1945-46 and the Cold War. *Political Science Quarterly*, 89(1), 117-146.

1942. Iran- USSR- Great Britain. *The American Journal of International Law*, 36(3), 175-179.

Jessup, N.H., 2001. Ambassadors and Chief of Mission to Iran.

Khadduri, M., 1951. Iran's Claim to the Sovereignty of Bahrayn. *The American Journal of International Law*, 45(4), 631-647.

Kibroglu, M., 2006. Good for the Shah, Banned for the Mullahs: The West and Iran's Quest for Nuclear Power. *Middle East Journal*, 60(2), 207-232.

Kinsella, D., 1994. Conflict in Context: Arms Transfers and Third World Rivalries during the Cold War. *American Journal of Political Science*, 38(3), 557-581.

Kurzman, C., 1995. Historiography of the Iranian Revolutionary Movement, 1977-79. *Iranian Studies*, 28(1), 25-38.

Kurzman, C., 1996. Structural Opportunity and Perceived Opportunity in Social-Movement Theory: The Iranian Revolution of 1979. *American Sociological Review*, 61(1), 153.

Ladjevardi, H., 1983. The Origins of U.S. Support for an Autocratic Iran. *International Journal of Middle Eastern Studies*, 15(2), 225-239.

Lambton, A.K., 1946. Some of the Problems Facing Persia. *International Affairs (Royal Institute of International Affairs 1944-)*, 22(2), 254.

Lambton, A.K., 1957. The Impact of the West on Persia. *International Affairs (Royal Institute of International Affairs 1944-)*, 33(1), 12.

Looney, R., 1988. The role of military expenditures in pre-revolutionary Iran's economic decline. *Iranian Studies*, 21(3), 52-83.

Lorentz, J.H., 2006. *Historical Dictionary of Iran* 2 ed., Lanham, MD: Scarecrow Press.

Love, K., 1953. Shah, Back in Iran, Wildly Acclaimed; Prestige at Peak. *The New York Times*.

Love, K., 1953. Shah Flees Iran After Move to Dismiss Mossadegh Fails. *New York Times*.

Marsh, S., 1998. The Special Relationship and the Anglo-Iranian Oil Crisis, 1950-4. *Review of International Studies*, 24(4), 529-544.

McLean, D., 1978. English Radicals, Russia, and the Fate of Persia 1907-1913. *The English Historical Review*, 93(367), 338-352.

Meekison, V.V., 1950. Treaty Provisions for the Inheritance of Personal Property. *The American Journal of International Law*, 44(2), 313.

Menon, R., 1982. The Soviet Union, the arms trade and the third world. *Europe-Asia Studies*, 34(3), 377-396.

Meskill, C.M., 1995. American Diplomacy in the Iranian Revolution, 1976-1981.

Millman, B., 1998. The Problem with Generals: Military Observers and the Origins of the Intervention in Russia and Persia, 1917-18. *Journal of Contemporary History*, 33(2), 291-320.

Mokhtari, F., 2005. No One Will Scratch My Back: Iranian Security Perceptions in Historical Context. *Middle East Journal*, 59(2), 209-229.

1952. New Iranian Chief Political Veteran. *The New York Times*.

Nikazmerad, N., 1980. A chronological survey of the Iranian revolution. *Iranian Studies*, 13(1), 327-368.

Pliskin, K., 1980. Camouflage, conspiracy, and collaborators: rumors of the revolution. *Iranian Studies*, 13(1), 55-81.

Pirouz, K., 2001. Iran's Oil Nationalization : Musaddiq at the United Nations and His Negotiations with George McGhee. *Comparative Studies of South Asia, Africa, and the Middle East*, 21(1/2), 110-117.

Pirouz, K., 2008. The Truman-Churchill Proposal to Resolve the Iran-U.K. Oil Nationalization Dispute. *Comparative Studies of South Asia, Africa and the Middle East*, 28(3), 487-494.

Pryor, L.M., 1978. Arms and the Shah. *Foreign Policy*, (31), 56-71.

Quosh, C., 2007. *American Foreign Policy Towards Iran: Between Values and Interests or Beyond?*, Hamburg.

Rahnema, S., 1990. Multinationals and Iranian Industry: 1957-1979. *Journal of Developing Areas*, 24(3), 293-310.

Razi, G.H., 1987. The Nexus of Legitimacy and Performance: The Lessons of the Iranian Revolution. *Comparative Politics*, 19(4), 453.

Renton, A.W., 1933. The Revolt Against the Capitulatory System. Journal of Comparative Legislation and International Law, Third Series, 15(4), 212-231.

Ronfeldt, D., 1978. Superclients and Superpowers: Cuba:Soviet Union/ Iran: United States, Santa Monica, CA.

Rosenthal, A., 1951. Britain-Iran Talk in U.N. Is Sought. The New York Times.

Ross, A., 1952. Hundreds Seized in Iranian Rioting Over Ghavam Rule. The New York Times.

Ross, A., 1952. Iranian Deputies Rebuff Mossadegh Over Martial Law. *The New York Times*.

Ross, A., 1952. Mossadegh Is Back as Premier of Iran ; Order Is Restored. *The New York Times*.

Ross, A., 1952. Mossadegh Out as Premier ; Ghavam to Take. *The New York Times*.

Rubin, B., 1980. American relations with the Islamic Republic of Iran, 1979-1981. *Iranian Studies*, 13(1), 307-326.

Rubin, M.A., 1995. Stumbling through the " Open Door ": The U.S. in Persia and the Standard-Sinclair Oil Dispute, 1920-1925. *Iranian Studies*, 28(3), 203-229.

Ruehsen, M.D., 1993. Operation 'Ajax' Revisited: Iran, 1953. *Middle Eastern Studies*, 29(3), 467-486.

1912. Russia and Persia. *The American Journal of International Law*, 6(1), 155-159. 1924. Treaty of Peace. *The American Journal of International Law*, 18(1), 4-53.

Saghaye-Biria, H., 2009. United States Propaganda in Iran: 1951-1953.

Sanjian, G.S., 1999. Promoting Stability or Instability? Arms Transfers and Regional Rivalries,1950-1991. *International Studies Quarterly*, 43(4), 641-670.

Shoamanesh, S.S., 2009. Iran's George Washington: Remembering and Preserving the Legacy of 1953. MIT International Review.

State, U.D., 1953. First Progress Report on Paragraph 5-a of NSC 136/1, "U.S. Policy Regarding the Present Situation in Iran", Washington, D.C.: State Department.

Stewart, I., 1926. American Treaty Provisions Relating to Consular Privileges and Immunities. The American Journal of International Law, 20(1), 81-102.

Stowell, E.C., 1924. The Imbrie Incident. The American Journal of International Law, 18(4), 768-774.

Sykes, P., 1921. South Persia and the Great War. *The Geographical Journal*, 58(2), 101-116.

Tapp, J., 1951. The Soviet-Persian Treaty of 1921. *The International Law Quarterly*, 4(4), 511-514.

The New York Times, 1953. Moscow Says U.S. Aided Shah's Coup. The New York Times.

The Associated Press, 1954. Mossadegh's Aide Seized in Teheran. The New York Times.

The Associated Press, 1954. Statements on Iran Oil Accord. The New York Times.

The New York Times, 1953. Reversal in Iran. The New York Times, 12-15.

The New York Times, 1953. Shah Instituted Iranian Reforms. The New York Times.

The New York Times, 1953. Shah Is Flying Home. The New York Times.

1907. The Recent Anglo-Russian Convention. *The American Journal of International Law*, 1(4), 979-984.

The United Press, 1953. Shah Leaves Rome to Fly to Teheran. *The New York Times*.

Times, T.N., 1952. World Court Bars Ruling on Iran Oil. *New York Times*.

United Press, 1949. Ruler of Iran Is Wounded Slightly by Two Bullets Fired by Assassin. *The New York Times*.

United States, 1918. Papers Relating to the Foreign Relations of the United States 1918: Supplement 2 The World War., 50(644).

United States, 1935. Papers Relating to the Foreign Relations of the United States, 1919. Volume II., 50(4).

United States, 1936. Papers relating to the Foreign Relations of the United States, 1921: Volume II., 44(3).

United States, 1938. Papers Relating to the Foreign Relations of the United States, 1923 Volume II. *Office*, II(397).

United States, 1939. Papers Relating to the Foreign Relations of the United States, 1924 Volume II. *Office*, II.

United States, 1940. Papers Relating to the Foreign Relations of the United States, 1925 Volume II. *Secretary*, II.

United States, 1943. Papers Relating to the Foreign Relations of the United States, 1927 (In Three Volumes) Volume III. *The American Historical Review*, 48(3).

United States, 1947. Foreign Relations of the United States, 1932 Volume II. *New York*, II.

United States, 1949. Discharge of Fiduciary Obligation to Iran., (1155).

United States, 1949. Discharge of Fiduciary Obligation To Iran (Report No. 1145). *Secretary*, 1-6.

United States, 1949. Mutual Defense Assistance Act of 1949.

United States, 1953. Foreign Relations of the United States, 1935 Volume I. *Policy*, I.

United States, 1953. Foreign Relations of the United States Diplomatic Papers, 1936 Volume III., III(369).

United States, 1954. Mutual Security Act of 1954. *Defense*.

United States, 1955. Foreign Relations of the United States, 1945. *Secretary*, (154).

United States, 1959. Report to Congress on the Mutual Security Program For the first half of Fiscal Year 1959. *Annals of botany*, (231).

United States, 1960. Foreign Relations of the United States, 1945 The Conference of Berlin (The Potsdam Conference)., II.

United States, 1960. *Foreign Relations of the United States Diplomatic Papers: The Conference of Berlin (The Potsdam Conference) 1945 (In Two Volumes) Volume I*, Washington, D.C.: GPO.

United States, 1961. Foreign Relations of the United States Diplomatic Papers The Conference at Cairo and Tehran, 1943.

Unknown, 1953. Fears of Tehran Merchants Concerning the Tudeh Party.

Unknown, 1953. *Proposed Course of Action with Respect to Iran*, Washington, D.C.: National Archives.

Whitney, P.D., 1953. Britain Is Cautious on Revolt in Iran. The New York Times.

Wilber, D.M., 1969. Overthrow of Premier Mossadeq of Iran: November 1952-August 1953, Langley, VA: CIA.

Wilson, K.M., 2002. Creative Accounting: The Place of Loans to Persia in the Commencement of the Negotiation of the Anglo-Russian Convention of 1907. *Middle Eastern Studies*, 38(2), 35-82.

Yegorova, N.I., 1996. *The "Iran Crisis" of 1945-46: A View from the Russian Archives*, Washington, D.C.

Zahrani, M.T., 2002. The Coup that Changed the Middle East: Mossadeq v. the CIA in Retrospect. *World Policy Journal*, 93-100.

Zirinsky, M.P., 1992. Imperial Power and Dictatorship : Britain and the Rise of Reza Shah, 1921-1926. *International Journal of Middle East Studies*, 24(4), 639-663.

Zirinsky, M.P., 1993. Render Therefore unto Caesar the Things Which Are Caesar's: American Presbyterian Educators and Reza Shah. *Iranian Studies*, 26(3), 337-356.

Chapter 5
Economic Relations between the United States and Iran

This chapter examines the economic relations between the United States and Iran. The economic relationship between Iran and the United States has been contentious from the beginning. Given the great geographical distance the beginning of diplomatic and economic relations was slow. Iran was willing to trade as long as it did not perceive it was being used as a pawn in great power politics or was being used as a source of cheap commodities. The United States considered its global position in relation to other powers and its interest in the region when dealing with Iran. The economic relations between the two nations tend to ebb and flow depending on the domestic politics of the time. The selectorate or the groups who select the leader in each nation and their preferences are reflected in the policies that each nation used in its trade relations with the other. For the United States the overall interests of big business that employed millions of American workers are seen in the quest for commodities, while in Iran income for the various governments was the primary motivation in trade agreements. Under Iran's authoritarian system the support of the elites who profited from trade were important to the various Shah's who sought elite support for their continued rule. More often than not, survival was more important for Iran than economic development, which was secondary until Iranian interests became more entrepreneurial. No matter what motivated American and Iranian leaders the economic relationship began as a rational political move but it has been overshadowed by the ongoing low-level of conflict between both nations that drive relations between the two.

As explained earlier, United States-Iran economic relations were very weak in the early days of contact. One of the earliest records of "official" analysis of trade potential dates back to Hampden Winston, the Minister Resident/Consul General of the United States in Iran. He presented his credentials on April 4, 1886 and left his post on June 10, 1886. He reported that there was no trade benefit or potential. His conclusions were based on the lack of property rights, the lack of political leadership, and the rapid and continuous decline in the value of Iranian currency. Nevertheless, towards the end of the nineteenth century some economic ties were developed between the two countries. During Spencer Pratt's (1866-1891) service at the United States embassy in Iran, he persuaded the President of the United States to consider trade between the two countries. Pratt also served as a representative of the Gatling Gun Company. Further, he obtained the concession to operate a power company for 60 years, which he sold to Francis Clercue (Mojani 1384 HS (2005 AD): 65). Numerous negotiations and contacts were underway until 1901 when the City of Buffalo invited the Iranian government to send an envoy to visit a construction expo to inspect equipment. Although, during this period, the United States was moving towards self-sufficiency, Clercue was eagerly exploring all possibilities for export. During this time, numerous agreements and contracts were drawn for the exchange of weapons, ships, and grain from the United States for wool and cotton from Iran. Other trade opportunities were increasing as well (Amerie 1925, Bonakdarian *nd*, Foran 1989, and Ghaneabassiri 2002). For example, there were over 20 large stores in New York selling Iranian rugs (Mojani 1384 HS (2005 AD): 68). Numerous requests from United States-based legal or individual entities to obtain concessions to establish banks, the right to mine minerals (especially oil), and to build railroads were reported. In 1911, Nabildoleh (Mojani 1384 HS (2005 AD): 70), the Consular of Iran in Washington D.C., contacted Standard Oil of New Jersey to investigate the possibility of oil exploration in Iran. He was responsible for congregating Iranian craftsmen from around the United States into Southern California and was instrumental in securing financial advisors from the United States.

Eventually, the United States began receiving requests from Iranian officials regarding the rate of exchange based on gold and silver prices, minting of coins, detailed information about Iranian markets and marketing means, a list of goods in demand in Iran, medical conditions, production of pharmaceutical products, and the regulations governing employment and trade by Americans in Iran. There is evidence that the Iranian government provided detailed information about these and other trade related information requests for the United States, in a sense helping the United States dominate Iran's trade and economy for years to come (Amerie 1925, Bonakdarian *nd*, Foran 1989, and Ghaneabassiri 2002).

Oil-Based Relations

The primary factor determining trade between the two countries was oil. In the tradition of the colonialism and the newfound power of the United States, relations with Iran regarding oil exploration were based on obtaining concessions-a standard tactic of the time. The first contact between Iran and the United States for oil exploration and concessions dates back to the first parliament immediately following the Constitutional Revolutions of 1907, during which a discussion regarding exchange of ships for oil occurred although never materialized (Foran 1989, Ghaneabassiri 2002, Paine & Schoenberger 1975, Volodarsky 1983). Numerous contacts were made between Iran and the United States exploring the possibility of establishing an American oil company in Iran to provide oil for the United States' fuel needs in Persian Gulf and Indian Ocean with myriad alternative financing schemes considered. However, as negotiations began to produce results and the Americans were requested to submit a proposal, Khostaria's concession was revealed. As a result, the Russians responded thus creating unrest in areas south of the Caspian Sea ending with the landing of Russian troops at the Port of

Anzaly. In 1911, the Russian occupation of the northern regions of Iran decisively put an end to all oil exploration discussions with the United States (Badakhshan & Najmabadi 2004, Daniel 2001, Foran 1989, Ghaneabassiri 2002, Greaves 1968, Kazemi 1985, Millspaugh 1933, Rubin 1995). The series of contacts began in earnest after the 1919 provisional agreement between Iran and Britain which brought both the United States and the Soviet Union in direct opposition with Britain. The United States and Russia were in agreement with Iran demanding that the agreement be nullified and Khostaria's oil concession, which had been purchased by Britain, be canceled (Badakhshan & Najmabadi 2004, Daniel 2001, Foran 1989, Ghaneabassiri 2002, Greaves 1968, Kazemi 1985, Millspaugh 1933, Rubin 1995). Later, however, the Soviet Union claimed that under provisions of existing treaty Britain could not legally obtain Khostaria's oil concession. In an effort to solidify oil relations, Iran's consulate was invited to attend the annual meeting of the American Petroleum Society in 1920 (Badakhshan & Najmabadi 2004, Daniel 2001, Foran 1989, Ghaneabassiri 2002, Greaves 1968, Kazemi 1985, Millspaugh 1933, Rubin 1995).

Between 1935 and 1939, the United States signed numerous trade agreements to limit trade with European countries. Between 1941 and 1946, the United States initiated the creation of the International Monetary Fund (IMF), the World Bank, and the International Bank for Reconstruction and Development. The role that these institutions played in making the United States the financial powerhouse became evident after World War II. Unlike during the nineteenth century, the United States no longer tried to secure its foothold in Latin America. Therefore, the U.S. and the Soviet Union did not wait for the war to end before creating spheres of influence, while also avoiding another stock market crash. One solution was through trade, which assured availability of (inexpensive) raw material for factories and suppliers.

As the only major undamaged country, World War II enabled the United States to become the economic leader of the world as industrial and agricultural capacities had actually become much more productive than in the prewar era. Although much of the increased capacity during the conflict was to aid the war effort, the US was able to divert production to non-military and consumer goods fairly simply. By the end of the war, two-thirds of the global commercial naval capacity belonged to the United States. This level of dominance was achieved in two ways: the expansion of the United States naval capacity, and the reduction of the naval capacity of the rest of the world. The economic might of the United States was complemented by a massive 12.5 million strong military, which boasted the world's only nuclear power. Furthermore, it had a $20 billion gold reserve. For a while the United States was the only superpower of the world. However, the Soviet Union was beginning to emerge as a potential rival. A major difference between the Soviet Union and other European countries was the ideological difference between socialism and capitalism. The Soviet Union showed its appetite for controlling other countries by erecting and supporting communist regimes in Eastern Europe, by maintaining its military presence in Iran, and trying to undermine the governments of Turkey and Greece. The areas of Central and East Asia that were given up in the dawn of socialist revolution were eventually incorporated. When the Soviet Union tested its first atomic bomb in 1949, the United States realized that it now had a serious and formidable rival. The Soviet Union's actions in Greece, along with its attempts at securing the straits through treaties with Turkey (suggesting a ban on passage by ships that were not registered in states bordering the Black Sea) necessitated that the United States begin a global opposition to the Soviet Union. Consequently, an all-out effort to contain the Soviets by creating a ring of anti-Soviet (or at least pro-America countries) around the Soviet Union was undertaken.Driven by great power rivalry, Iran saw the United States as an alternative to British hegemony and Soviet expansionism, while the United States saw Iran as a way to gain a foothold in the region and frustrate the Soviet Union and British. The end result was a good but wary relationship.

Post World War II

The period immediately following World War II was one of turmoil. On one hand, Iran was trying to free itself from the occupying forces of Britain, the Soviet Union, and the United States. On the other hand, it was trying to overcome economic problems by selling oil, borrowing, or securing aid from mostly the same countries or their allies. Even before the end of WWII, Americans were actively seeking oil concessions from Iran. In February of 1944, Socony Vacuum and Sinclair Oil contacted Iran to obtain concessions for exploration of oil in Baluchistan in southeast Iran. The hope was that oil explorations away from the five northern states and away from the southwest-central part of the country would not invoke Soviet or British reaction or retaliation. Somehow, Royal Dutch Shell (a British company) managed to have a similar offer on the table on the same day. One can only imagine the difficulties of negotiating with the two most powerful nations in the world, one on the decline and the other on the rise. While the United States and Britain were jostling for position, they worked with Iran to keep all negotiations secret from the Soviets (Badakhshan & Najmabadi 2004, Daniel 2001, Galpern 2002, Kazemi 1985, Marsh 1998, Pirouz 2001, Rubin 1995). Nevertheless, on September 1944, the Soviet Union also proposed an oil concession. Eventually, Prime Minister Saed rejected all three (3) offers on October 18 and consequently lost his position. Mosaddeq convinced the Majlis to pass a law banning any concessions for as long as the country was under the occupation of foreign troops. Any Prime Minister that violated that law was to be prosecuted (Badakhshan & Najmabadi 2004, Kazemi 1985, Pirouz 2001).

After the war, the main plan of action for the Soviet Union and the United States was to create "friendly" and "sympathetic" governments around the world. In this regard, the Soviets continued their occupation of Iran after WWII ended, and they helped the leftists create autonomous governments in Eastern Europe, which were replicated in Azerbaijan and Kurdistan. Meanwhile, the Americans supported many authoritarians and dictatorial leaders such as Mohammad Reza Pahlavi. Mohammad Reza Pahlavi was brought to power by the occupying forces on September 16, 1941. Mohammad Reza Pahlavi's power grab also led to the exile of his father (Abrahamian 2008, Daniel 2001, Davis 2006, Ladjevardi 1983, Pollack 2005, Ronfeldt 1978). Obviously, for the US, the degree of attention and allocation of resources was proportionate to the degree of threat in a given country and the probability of its switching from one camp to the other. Initially, the two countries focused on Greece and Turkey, and then, they faced off in Iran. In Iran, the Democrat Party of Azerbaijan and the Komeleh Party of Kurdistan declared their autonomy from the central government with the help of the Soviet Union (Abrahamian 1970, 1978, 2008, Behrooz 2001, Daniel 2001). The Soviets also secured an oil concession before withdrawing their troops from Iran. The United States' response was to provide economic aid to its allies and friends either directly or through the United Nations Relief and Rehabilitation Administration (UNRRA). For example: in February 1947, when Britain announced that it could not hold the guerillas in Greece, the United States gave Greece and Turkey $300 million in aid.

At first, the United States was slow to support Iran. After the war, referring to the Tehran Agreement (1943), Iran demanded payments for war damage and other economic compensation. However, the United States was reluctant to pay, and it accused Iran of inflating war damage (Jafari Valdani 1382 HS (2004 AD): 24), even claiming no damage to the natural resources of Iran (Jafari Valdani 1382 HS (2004 AD): 53). It seems that the attention, and thus the help from the United States to Iran, was directly related to the conduct of the Soviet Union in Iran and the level of domestic unrest. At least initially, most of the United States' assistance to Iran was in the form of weapons. The above mentioned source also acknowledged that the United States military aid created tension between Iran and the Soviet Union (Bonakdarian *nd*, Gavin 1999, Pryor 1978, Rick 1979, Ronfeldt 1978, Seitz 1980).

While the Soviet troops were in Iran, Prime Minister Qavam feigned to be leaning to the East, as reflected through the inclusion of the Tudeh Party members in his cabinet. With the aid of the United States Qavam was arming and equipping the military. Once the army was strong enough, Qavam demanded that the army should be present at all polling places for the upcoming Majlis elections, including polling places in Azerbaijan. When the Tudeh Party and Democratic Party of Azerbaijan objected, the army moved into both regions and defeated the communist supporters' bases in both Azerbaijan and Kurdistan. By June 8, 1947, the United States had given Iran a $10 million loan to purchase weapons and other military equipment, which was augmented on the following October to provide additional advisors and technicians (Abrahamian 1970, 1978, 2008, Behrooz 2001, Daniel 2001).

Within the few months following Qavam's government, Iran had three Prime Ministers. The United States' concern about Iran's instability gave cause for providing additional advisors and technicians to Iran. In addition, another $25 million aid package was allocated. Within the next months, after another attempt on the Shah's life, martial law was declared and the communist party of Tudeh was declared illegal.

Point IV Program

In January 1949, Truman revealed his Point IV program. Although the official reason for Point IV was to help poor countries through technical support, its main objective was to halt or slow the expansion of communism around the world. In November of 1949, the Shah traveled to the United States to secure even more aid and military equipment. He pointed out that he had sold his "ancestral" land to the peasants at low prices. Iran was the first recipient of Point IV funds in the Middle East. Soon, the United States provided aid and loans to Iran to improve its economy. Although American weapons, technicians, loans, and other forms of aid helped Iran and the Shah's regime, United States-Iran relations remained uneasy, as they could trigger a military reaction from Moscow (Amuzegar 1958, Harris 1953, Ricks 1979, Seitz 1980, Summitt 2004). Although aid from the United States to Iran increased substantially from 1946 to 1952 (from $3.3 to $44.1 million) Iran was not as effective as Greece or Turkey in securing aid from the United States. For example, in 1946, the United States aid to Turkey and Greece was $6.1 and $195.2 million dollars, respectively. But, by 1952, it was $259.0 and $351.2 million, respectively. At the time, Turkey had a population size comparable to that of Iran, while Greece's population was much less. The land area of Turkey is about half that of Iran. Greece is much smaller than either country. Attributing the difference in aid to strategic importance of the three countries would be a mistake. Iran had access to oil revenue and the United States would have been wise to allocate its resources as it did then. In fact, the United States decided to redirect its aid to Iran through Point IV and to use it as leverage to promote economic development and social changes. Therefore, Mosaddeq's suggested 1952 land reform was well received by supporters of economic aid such as the aid tied to Point IV (Amuzegar 1958, Ricks 1979). In general, the United States liked Mosaddeq's agenda because it was in line with US policies. However, he was independent, opposed the Shah, and could work with the communists. In addition, Mosaddeq wanted to keep Iran neutral. He knew that Iran needed aid from the US and supported Point IV, but he did not like the strings that were attached. In the 1950s, the United States demanded three (3) conditions in return for its aid: 1) anti-Soviet and anti-China foreign policy, 2) permission to have foreign investment (especially from the United States), support for investors, and protection of investors, and 3) internal stability, which was interpreted as destruction of the leftist and communist movements. In 1952, Mosaddeq supported a $23 million "technical" support loan under Point IV and appointed a high-ranking Iranian team to negotiate terms with the United States. However, when the United States agreed to give the loan under the "Mutual Security Act" without an official treaty under the Point IV umbrella, Mosaddeq opposed it because it would have violated Iran's neutrality (Amuzegar 1958, Ricks 1979). Since Iran was still contesting the terms of the Anglo-Persia Oil Company's concession and was trying to nationalize the oil, there was hardly any revenue and Iran was in desperate need of foreign currency (Abrahamian 2001, 2008, Dadkhah 2001, Daniel 2001, Davis 2006, Marsh 1998, Paine & Schoenberger 1975, Young 1962). It was apparent that Mosaddeq was independently minded and would not be influenced by the United States. The United States could not take a chance without him either. At the time, as explained earlier, the communist party of Tudeh had great organization and could fill the vacuum created by the collapse of Mosaddeq's regime. This would have created yet another Soviet ally; one with access to warm waters, vast oil reserves, and the ability to block the Hormuz strait (effectively blocking the flow of oil from Iran, Kuwait, Iraq, and the United Arab Emirates). The election of a Republican president in the United State ended the remaining economic aid to Mosaddeq's government. The United States went further to join England and arrange for a *coup d'état* to completely remove Mosaddeq and shore up the Shah and put Iran in the United States' camp once and for all (Abrahamian 2001, Gavin 1999, Ladjevardi 1983, Marsh 1998, Pryor 1978, Ronfeldt 1978, Shoamaesh 2009, Young 1962, Zahrani 2002). Removal of Mosaddeq paved the road to pour money, equipment, advisers, and above all, weapons into Iran. Members of the Tudeh Party were arrested and prosecuted, and their supporters were harassed. As history has revealed, the danger of communism in Iran evaporated.

The Point IV program can be seen as a reflection of the times. The United States wanted to contain the Soviet Union and Iran was seen as a key player in the containment strategy. Americans felt that an economically developed Iran would have a substantial middle class that would support the Shah and his development programs. A strong middle class that had disposable income, education, and some luxury goods was seen as a bulwark against communist influences. The Shah saw American aid as good for his rule and development of Iranian economy. Leadership in both nations, however, erred in thinking that the Iranian people could be lead down the road of economic development by outside influences. The roots of the revolution stretch back to the economic relationship with the West and the United States in particular. The removal of a popular prime minister further exacerbated the simmering resentment. While reflective of the political realities in American politics and the Iranian court the perceptions of policy makers failed to take the great mass of the Iranian population into account when development plans and political maneuvers were carried out.

After the removal of Mosaddeq, the Secretary of State offered to give $30-35 million in military aid, $25 million through Point IV, and another $60 million to help with the budget, provided that the United States Congress approved. On September 1, 1953, the office of Point IV informed Prime Minister Zahedy that based on agreements between the United States and Iran; they could provide $23.4 million for economic development. Of this amount, $20.4 million was for existing programs and the rest was to be used for new programs (Amuzegar 1958, Ricks

1979). Although in 1954 Eisenhower recommended $60 million as emergency aid to Iran, he favored long-term loans. Nevertheless, the United States agreed to continue $23 million in aid per year for the near future. Later, Iran benefited from the aid to the Middle East, which was promoted to keep the Soviet Union out of the region. By 1956, there were numerous agreements on military and economic cooperation between the two countries and 45% of the United States' loans to Iran were for military purposes. In this year, the United States eliminated aid to Iran's budget, reduced economic aid, and increased military aid (Amuzegar 1958, Ricks 1978). Unfortunately, the 1958 revolution in Iran again caused much unease for the United States.

The way Point IV Operated

The operation of Point IV in Iran required the establishment of special offices in Iran and required authorization from the Majlis and the government (Amuzegar 1958, Harris 1953, Seitz 1980). The plan for economic development and welfare improvement under this program was implemented in many ways.

A joint commission for growth and development of rural areas was based on a bill to increase the peasant's share of output and its amendment, which established the National Growth Foundation. The bill demanded that 20% of agricultural and rural real estate must be used to improve the welfare of the peasants. Half of these funds were to be paid to the farmers and the other half were to be used for building and maintenance of bridges, schools, mosques, public baths, infirmaries, low income housing, and drinking water. The Ministry of Internal Affairs was to establish an office to oversee the implementation of the bill and the Ministries of Internal Affairs, Budget, Agriculture, Education, Health, Transportation, as well as the Planning Organization, National Bank, and the Agriculture Bank were responsible to implement the bill (Amuzegar 1958, Clapp 1957, Harris 1953, Kristianson 1960). On June 28, 1953, an agreement between the two countries was signed and a joint fund for development was established. The director of this fund was appointed by the prime minister and each of the above-mentioned ministries provided a liaison to attend to daily routines of the new office and provide feedback (Amuzegar 1958, Seitz 1980). Other subsidiary commissions were created which were operated by one of the officers of the International Cooperation Administration (ICA) and the governors of the provinces in Iran. These directors were responsible for coordinating and implementing different projects in their respective regions. Inspectors of the ICA and an Iranian representative were to verify appropriate performance of the projects. The ICA group consisted of nine ancillary branches of education, social development, agriculture, labor, education, industry, engineering, construction, and audio-visual (which were supported by office staff) (Amuzegar 1958, Seitz 1980). As of June 30, 1955, there were 312 American and 106 Iranian staff assigned to the project.

The American officials insisted that the United States personnel who were working under Point IV be exempt from the customs and other taxes. This had become a common practice for the United States operations in other countries. Other recipients of the United States aid were treating aid workers as members of the United States diplomatic envoy and giving them diplomatic protection and tax exemptions. In a letter dated January 19, 1952, William Warren, the director of Point IV from 1952 to 1955, asked for similar privileges from Prime Minister Mosaddeq (William 1999). He responded that:

> The government of Iran agrees that upon approval of the government of the United States a technical team from that country to take the responsibility of conducting the necessary duties and responsibilities for conducting technical cooperation between two countries and the government of Iran will do everything possible within Iranian Constitution and other existing laws and regulations to consider those workers as part of the political envoy of the United States in Iran and extend the privileges and immunities of political envoy, in accordance to their rank. The Iranian government agrees to be responsible for all the taxes, customs, and duties that apply to this especial group or each of their workers or their accompanying family members.

The Americans used this letter as proof of immunity for all United States citizens and as evidence that they were not subject to the Iranian judicial system. However, Iran did not agree, and as it was explained earlier, on October 3, 1964 the Majlis had to pass a bill that legally granted diplomatic immunity to all United States workers in Iran. Later, this issue was revealed by Khomeini, set off the historic 15 Khordad uprising, Khomeini's exile, and eventually the Islamic revolution of 1979 (Mahdavy 1965, Pollack 2005, Summitt 2004).

It is worth mentioning that by 1953, the Tudeh Party had begun its opposition to the Point IV program, and in some parts of the country, there were objections to the program by local people. This caused worries among the Americans about the safety of its personnel and offices. Also, as a result of the opposition, not all of the funds were disbursed. In the same year, Iranian and American officials signed a five-year agreement on rural education and development.

There were numerous organizations involved in implementing Point IV, both governmental (American as well as Iranian) and private. The Near East Foundation's objectives overlapped with those of the Point IV program in the areas of education and rural development. Therefore, they cooperated in these areas. Eventually, they signed a cooperation agreement on January 13, 1954.

On April 23, 1955, the Point IV program was moved to the Secretary of State's office and became the office of International Cooperation Administration (ICA). On June 30, 1955, the Office of Foreign Aid was closed in the United States and all the military aid programs were transferred to the Department of Defense (Azumegar 1958, Ricks 1979). However, this did not the end the program. The program actually lasted until 1967 and resulted in 162 development projects encompassing health, agriculture, industry, and education. According to Azghandi (1376 HS (1997 AD), until 1961, 50 percent of the rural development budget was provided by the Point IV program. In the National Security Council (NSC) report 129/1 dated April 24, 1952 referenced by Hamraz (1381 HS (2002 AD) the main concern of the NSC was to halt economic and social turmoil around the world, to prohibit the emergence of countries that leaned towards the Soviet Union, and to minimize danger to the West. Hamraz (1381 HS (2002 AD) adds that the United States' economic assistance was geared to protect the political and economic interests of the West. During this period, third world countries were following the United States' recipe for development based on economic planning, and at least in the minds of American experts such as Rostov and Millikan, it was obvious that third world countries would be better off producing agricultural products and extracting minerals to obtain foreign currency. In turn, these countries could use foreign currency gained to buy manufactured goods produced in the West, made from the very same primary products that the third world countries produced (Hamraz 1381 HS (2002 AD)). Simultaneously, the United States was busy spreading its style of education as well as training third world elites to create an anti-socialist, pro-west atmosphere in these countries (Carnoy 1974). Iran also initiated its first "7-year development plan" in 1946. However, it was not implemented due to financial and other exigencies. The plan was revised in 1948 and financed via oil revenue, borrowing from the National Bank (Bank Melli), a loan from International Bank for Reconstruction and Development, and loans from domestic and foreign companies (Araghi 1989, Daftary 1988, Ricks 1979). Iran's first development plan came to a halt due to lack of funds caused by a British blockade of Iranian oil in response to Mosaddeq's attempts to nationalize the oil industry. From 1950 until the *coup d'état* against him, aid from the United States was Iran's only source of foreign exchange. Although aid from the United States was vital for Iran's survival, it was unreliable. The United States was involved in a delicate balancing act. The US wanted to stabilize Iran to avoid the spread of communism, but did not want to support Iran to the point that it would not need the oil revenue to postpone or derail oil nationalization. Oil nationalization was important for several reasons. It would cause direct financial damage for England, as well as direct and indirect loss to other western countries involved in oil exploration and extraction such as the United States. The main fear was a domino reaction of oil nationalization. Although the United States wanted to help its ally, it also was trying to replace England's presence around the world.

Aid Impact on Iranian Agriculture

The Point IV program provides a good example of how the best intentions ended up hurting the Iran-US relationship more than helping it. At least during the middle of the twentieth century, the common belief was that whatever worked in the United State should work everywhere else. For example, mechanized agriculture made the United States a major exporter of agricultural products. Therefore, mechanized agriculture is better than other alternatives. This mentality was based on a lack of understanding by the experts of the era. These experts ignored all the other agriculture differences that existed between the two nations, and only concentrated on a mechanized approach. In the case of Iran, many other factors differentiated the two countries. Iran was a country with little water and rainfall, small farms, more rugged landscape, little or no infrastructure such as roads, and an owner-worker tenant system (Carey & Carey 1976, Connell 1974, Freivalds 1972, Kristianson 1960).

American experts concluded that traditional seeds were inferior to newer hybrids, so they began a plan to replace domestic planting seeds with hybrids. They also promoted land reform, which was becoming very popular. Furthermore, they promoted new mechanized farming (Carey & Carey 1976, Connell 1974, Freivalds 1972, Kristianson 1960). Eventually these methods proved damaging and costly for different reasons and many of the ideas were abandoned worldwide. For example, the new hybrid seeds were not quite right for other countries' environments and were not as resistant to draught, disease, and insects of the region. In some cases the entire crop of a region was decimated when diseases such as wheat rust attacked, such as in India and Iran.

Converting small plots of land to large, plantation style farms also proved harmful. Removal of traditional fences (which harbored insects, birds, and other small animals) destroyed the natural balance of the environment that had been established long ago. The result was unchecked growth of some harmful insects. This necessitated increased use of pesticides, which negatively impacted the environment. The increased use of pesticides and farm machinery, either by choice or by force, exacerbated financial needs because none of the machinery or chemicals was produced domestically (Carey & Carey 1976, Connell 1974, Freivalds 1972, and Kristianson 1960). This was not the case in the United States which evolved on its own and developed machinery when necessary.

Another reckless act was the drilling of numerous deep wells around the country. These wells depleted the underground reservoirs which not only destroyed the newly established mechanized farms, but also devastated the livelihood of farmers downhill from the reservoirs that depended on them for drinking and farming (Carey & Carey 1976, Connell 1974, Freivalds 1972, Kristianson 1960). What these advisors did not take into consideration was that one cannot use more water than the total precipitation of a region, regardless of how water is used. As a result of lack of water and ability to maintain and protect the fields, fertile lands were eroded and the nation's productive capacity suffered.

Another example of unexpected consequences of short-sighted agricultural recommendations was the destruction of domestic stock. Under a program to improve domestic poultry stock, farmers could receive 12 American chicks provided that they relinquish 12 domestic chicks in return. Consequently, domestic stocks of birds were rapidly replaced with imported ones. Upon the recommendations of the American advisors, these chicks were to be raised in cages. These required feed, vaccination, and medical care, none of which were available then and required importation (Carey & Carey 1976, Connell 1974, Freivalds 1972, Kristianson 1960).

The Shah's regime, the American government, and American private businesses benefited from opportunities created by these actions and changes (Carey & Carey 1976, Connell 1974, Freivalds 1972, Kristianson 1960). American businesses were pleased because another market for their products was created; the American government was happy thinking that these actions would result in loyalty and support for the American government and acceptance of American values (thereby stopping the progress of communism). The Shah's relatives and the "1,000 families" were making fortunes because they had a virtual monopoly in imports. The Shah and the regime were happy because they could strengthen their stranglehold on the country and its economy. Quietly, however, the seeds of discontent and eventually hatred towards the Shah and the United States were sprouting.

Another factor that contributed to the eventual outright hostility between the two countries was the fact that Americans were unwilling to adapt to the culture and customs of Iran. There was no attempt to accept or adopt Iranian bureaucracy, to work in tandem with Iranian institutions, or the citizens. Instead, efforts were made to replace rules, customs, cultures, and methods that existed (regardless of their importance or capabilities) with American ways, methods, and customs (Embry 2003, Seitz 1980). There could have been, and in some cases were, benefits in adopting different methods or learning a new way of doing things. The problem was the attitude and the mentality of superiority, which bordered on arrogance. Iranians, with their special blend of ancient Asian culture intertwined with Shi'a teachings and Sufi mentality did not help the cause. Furthermore, the declining power of the country over the previous 150-200 years helped to augment the mentality that anything Iranian was sub-standard and inferior to those of Western counterparts, especially to those of Americans (who not only managed to defeat England in 1776 and in less than 150 years become the most powerful and richest country in the world). This love-hate relationship was bound to end, sooner or later. Another negative factor was the feeling that Americans were using Iran to advance their own cause and fortune. The relationship was not based on mutual advantage, let alone on mutual respect. For example, the Americans stated that they would like to help Iran improve its agriculture using new techniques and tools. One of the provisions of the Point IV program was that only American technicians and consultants could be used and the American embassy made it clear that experts and volunteers from other countries such as Germany and Sweden were not allowed to participate in Point IV projects. The extent of involvement of other nationals must be limited to the projects that were conducted between other countries and Iran jointly (Hamraz 1381 HS (2002 AD)). Eventually, the seemingly unlimited support of the United States for the ever-increasingly unpopular Shah tilted the balance from love to hate.

Numerous other activities were occurring in conjunction with, as well as independent of, the Point IV project. One example was the activities of the consultants from the University of Utah, which began an agricultural promotion program in Tabriz and some other parts of the country in 1952. The following year, members of Iran's Ministry of Agriculture staff attended a follow-up conference in Beirut. Another program used widely was the 4H Youth Development Program, which was promoted in all countries that received Point IV aid (Embry 2003).

It seems that every effort was made, deliberately or unintentionally, to make the United States aid recipient countries depend ever-increasingly on the United States. For example, in 1951 when Iran requested a $25 million loan from Export-Import Bank of the United States (of which $6 million was to be used to purchase agricultural machinery) it received $53 million instead. However, no attempts were made to provide training to produce the necessary parts in Iran; instead, they were to be imported from the United States with no assistance from the United States government. Of course, a side-effect of mechanization of agriculture was to create many unemployed peasants. The situation got even worse after the so called "White Revolution" which exempted mechanized farms from land reform (Embry 2003, Freivalds 1972, Mahdavy 1965). The consequences of this sort of top-down reform of the agricultural sector was resentment on the part of the small farmers and even landlords. The agricultural sector of the economy did not expand as advertised and the resentment of the rural population increased. The Iranian revolution was primarily an urban revolution. However, the dissatisfied rural populace provided vital support when mobilized by the radicals.

Infrastructure

Heavy use of transportation facilities during World War II by the Allies caused substantial damage to the meager infrastructure of Iran. The damage was more substantial on the railroads, which were both over-used and improperly maintained. Part of the United States economic aid was devoted to refurbishing the country's railroad system (Abrahamian 2008, Bonakdarian nd, Clapp 1957, Dadkhah 2001, Daniel 2001, Gavin 1999, Ricks 1979). Here too, the main object was to meet the actual and potential needs of the United States. As such, the very first work connected the cities of Mashhad and Tabriz to Tehran, connecting the eastern part of the country to the western part in a line parallel to the Soviet Union's boarder. The rails were actually extended to the border cities of each region as well. These railroads were completed in 1956 and 1957, respectively. The maintenance cost of these railroads became a burden to the Iranian government. Since the public could not afford the required maintenance, they were in poor repair. Furthermore, airport development occurred in strategic cities such as Tabriz, Mashhad, and Khoramshas even though only a few could afford to fly to the Unites States, such as friends and families of the Shah and members of the 1,000 ruling families. Without a doubt, Iran had major deficiencies in all aspects of transportation; however, construction of this infrastructure was not designed to improve Iran's ability to grow. The infrastructure was built to help the United States contain the

Soviet Union. In the early years after WWII, the lack of domestic revenue or foreign exchange left only one alternative, to turn to the United States for financial assistance (Abrahamian 2001, 2008, Dadkhah 2001, Daniel 2001, Davis 2006, Marsh 1998, Paine & Schoenberger 1975, Young 1962). Transportation deficiency was only one problem facing the country, which was competing with myriad of other needs, including national defense. Thus, it was not as unreasonable that the United States wanted to spend its money where it benefited the United States' interests the most. The only problem is that much of the "American money" was actually "loans" that Iran had to pay back with interest.

Different types of American aid and loans were used in the late 1940s and early 1950s, especially under the Point IV program, to fund (in part or whole) different economic and infrastructure projects. The projects included cement, textile, canned fish, and sugar factories; dams and water treatment plants; as well as diverse investments to improve different industries from leather tanning to china production (Amuzegar 1958, 1992, Ansari 2001, Carey & Carey 1976, Clapp 1957, Freivalds 1972, Hetherington 1982, Karshenas & Hakimian 2004, Mahdavy 1965, Ramazani 1974, 1990). Iran benefited from additional resources provided by American assistance, however, the gradual influence of Americans and their attempt to change the structure of Iranian society created tension and friction. In the early days post World War II, the mere presence of Americans was welcome news to quell concern of the power and pressure of Soviet Union and England. Their financial assistance, even in the form of interest bearing loans, was the added gravy. As England lost its might, the danger of occupation by Soviet Union receded, and the communists were decimated by the Shah, the negative aspects of reliance on yet another foreign power were gradually becoming evident.

The hand of the United States was into everything in Iran and impacted all aspects of Iran's existence from economics, to politics, to social and cultural matters. Of course interaction of different forces in a society always plays a major role on active or inactive responses towards issues. For example, as early as 1949 a senator accused Americans of meddling in Iranian affairs; in 1950 the *New York Times* was worried that the presence of the United States in Iran might trigger the same negative sentiment as prevailed against the British in Iran; Harriman's visit that year did not help the cause either; in 1951 some of Point IV facilities were attacked; in 1952 Americans were accused of spying and Warren, the director of Point IV reported that almost everywhere that Americans visited they were confronted by "Yankee Go Home" slogans. In 1953, especially wherever the communists and socialists were strong, there were demonstrations in factories and in villages against American presence. In the same years, some 30 Point IV workers in the state of Fars had to be protected by armed tribal forces and had to be taken to their strongholds for protection; there were demonstrations by university students and many attacks against Americans. Needless to say the American *coup d'état* against Mosaddeq, declaration of Martial Law, the pursuing "cleansing of communists," and the establishment of a pro-American dictatorship put an end to all the visible and open opposition to the presence and influence of the Americans. At the same time these circumstances deepened abhorrence towards Americans.

During the 1950s the United States promoted economic development in countries friendly to the United States as a way of preventing the spread of communism. This decade also was the decade of accepting government intervention in economic affairs, especially with regard to development efforts in Third World countries. After World War II, it was apparent that planned economies can and do grow rapidly, as demonstrated by the Soviet Union. This had great appeal for Third World countries that were struggling to avoid falling further behind, while simultaneously aspiring for economic development and the pursuant standard of living. Using evidence based on the declining terms of trade and structural realities of those developing countries, Third World economists such as Prebisch were stating that dealing with the developed countries of the West was the main cause of Third World country demise. Communism was spreading, and the developing nations were distancing themselves from the West. The American solution was to establish dictatorships, such as the Shah's regime, in order to stop the avalanche of communism and begin a planned economic development program using Western values and economic models to either stop or, at least, to slow down the spread of communism (Amuzegar 1958, Ladjevardi 1983, Pryor 1978, Ronfeldt 1978, Zahrani 2002).

Starting with this period, almost all countries in the world began planned economic development of one kind or the other. Five and seven-year economic development plans that featured popular economic development models proposed by Lewis (1954) and Harrod-Domar were spreading everywhere. Both of these models were based on the (assumption) of shortage of capital and abundance of labor in Third World countries. The remedy to underdevelopment was to increase capital, which was generously offered by the United States. An example of the use of these models to overcome a lack of development with the aid of American funding is Ghana. Ghana was the first sub-Saharan country to gain its independence from colonial powers. The evidence suggests that it did not work in Ghana.

After a 1953 *coup d'état* in Iran, the Shah began consolidating his stranglehold on the country; first with financing, military assistance and advisors from the United States, and then later, mostly on his own, but still with support and approval from the United States (Ansari 2001, Mahdavy 1965, Pollack 2005, Ramazani 1974, Seitz 1980, Summitt 2004). Until 1960, the emphasis of the development plan was promoting agriculture, as this was the model that the West preferred. The West was to produce industrial goods; Third World countries were to produce agricultural, mineral, and primary goods. The respective surpluses were to be traded. During this period light industrial production, mostly in the form of parts assembly, began in most third world countries. The distinguishing mark of this period is the increase in importation of consumer goods to the point that, by 1958, imports exceeded exports. Iran became more dependent on the West without being able to utilize its own oil revenues to exit the group of underdeveloped countries.

The end of the decade, as explained previously, brought a new wave of instability to Iran. People were unhappy with the economic and social structure the United States was demanding social and economic changes to reduce the risk of communism. Under pressure, the Shah introduced his "White Revolution," which not only brought no economic improvement or social and political stability, but also created great opposition from landlords and religious leaders, especially Ayatollah Khomeini. Combined with legalization of the capitulary law, the opposition was intensified and most likely planted the seeds of revolution which bore fruit in 1979. Nevertheless, the Shah survived and used the outcome as leverage to become even more powerful, which in turn allowed him to become an outright dictator, as explained earlier.

The year 1973 and the Arab Oil Embargo mark another landmark for the region and hence Iran. During the period leading to 1973, Iranian oil revenues began improving, in part due to higher prices, a greater share of revenues, and larger production. During this period, some consumer goods (especially automobiles) were assembled in the country usually with poor quality, as was the case with most similar import substitution products in the majority of Third World countries (Alizadeh 2006, Floor 2005, Hetherington 1982, Karshenas & Hakimian 2004, Nowshirvani 2010, Pesaran 1997, Rahnema 1990). The intent was to create a sense of pride and modernization in the public. Instead, they employed wasteful ways of producing substandard products. Their main contribution to the national economy, however, was to make a few investors, mostly related to the court, unbelievably rich. The import substitution policies also upset major importers in the Bazaar, who either missed the opportunity to take advantage of high tariffs to get even richer, were deliberately left out because they had fallen out of favor with the Court, or had supported the opposition-especially the religious groups (Alizadeh 2006, Floor 2005, Hetherington 1982, Karshenas & Hakimian 2004, Nowshirvani 2010, Pesaran 1997, Rahnema 1990).

By 1973, the occurrence of several, mostly independent events, changed the course of history in Iran. By the end of 1960, the British lion was a shadow of its former glory days. England was withdrawing its military from almost every corner of the world, while the United States was trying to fill the gap, usually in a less rapid manner. The task had become even harder since the United States was still bogged down in other parts of the world: trying to fill the voids created by France, Belgium, Italy, Germany, and England. Most of these and other colonial powers, even those that were victorious in World War II, were losing their grip on their colonies. The colonies were declaring their independence left and right with or without the help of the Eastern Bloc and their allies around the world. The single biggest headache for the United States was its engagement in Vietnam and by proximity, the rest of South East Asia. In 1973 the Arab Oil Embargo gave the Shah an opportunity to really show its subservience to the West by increasing Iran's oil production to fill as much of the shortfall as possible.

It is not clear who came up with the idea that Iran could also help suppress regional unrest without getting the United States involved. This became a common practice as both the United States and the Soviet Union used their allies to stage wars on their behalf. Chapter 6 details Iran's military engagements with the United States around this period. Consequently, Iran did not need any further foreign aid or loans, American or otherwise. Iran actually found itself in the situation of not knowing what to do with all the money made possible from a larger share in OPEC and higher prices. The Shah was like a little boy in a toy store, buying all kinds of weapons, most of which were never delivered. Of the ones that were delivered, the more sophisticated arms remained in the hands of American technicians and advisors, even though they were paid for by Iran and were stationed in Iran. The rest of the money was used to suddenly increase the import of consumer goods, including luxury goods (Amuzegar 1992, Looney 1988, Mahdavy 1965, Nowshirvani 2010, Pesaran 1997, Pryor 1978, Rahnema 1990, Summitt 2004).

The regime did not know how to spend all that money. Prime Minister Hoveida bragged that "now we can send our dirty clothes to Europe on planes to laundry". There is no definite historical evidence that such an act of arrogance ever took place, but some Iranians were convinced that such wasteful things were possible and actually did happen. Suddenly, Tehran was like a big supermarket. Money was left to start buying bankrupt European factories that could no longer compete in the global market such as the German steel company Krupp Huttenwerke in 1974. By 1975, Iran was a major lending country in the world. It signed numerous bilateral agreements to invest billions of dollars in Italy, Germany, England, India, France, and numerous underdeveloped countries. None of these investments were returned to Iran after the revolution and it is not clear if Iran ever received any dividends from these investments. Within the country, the five-year development plan was scrapped and its goals were doubled.

Revolution and Economy

In 1979, Iran had a political, not an economic, revolution. Before the revolution, Iran had a typical rentier economy. The primary source of revenue for the country was oil. Massive foreign currency from a single source financed almost all activities. The majority of the country's needs were imported as is the case in such economies. The country suffered from the classical Dutch Disease syndrome. As the price and demand for oil increased, Iran's foreign exchange grew as well. Some of the revenue was siphoned back to the West, especially to the United States, to purchase weapons. The rest was channeled to the importation of everything from ketchup to luxury cars (Amuzegar 1992, Looney 1988, Mahdavy 1965, Nowshirvani 2010, Pesaran 1997, Pryor 1978, Rahnema 1990, Summitt 2004). The side-effect of Dutch Disease is currency overvaluation, which makes the price of foreign goods relatively cheap. Consequently, domestic producers are unable to compete with foreign goods. Due to heavy dependence on the United States dollar, the government adopted a "stable dollar price" policy. Gradually, domestic production lost its competitive edge and collapsed. In this economy, only the importers and industrial or agricultural producers that were subsidized, or were protected by tariffs, could and did prosper. Needless to say, many of these individuals were either part of the royal family or closely tied to the royal court. The majority of the importers were concentrated in the Bazaars of major cities, especially Tehran.

After the revolution, a vacuum was created by the departure of the royal family, their cronies, and the so called 1,000 families. These affluent people left the country and took their money with them. It is important to recognize that a certain level of chaos and disequilibrium occurs in any social unrest, especially a revolution. In the case of Iran, several other factors augmented the problem. The first was that the main leaders of the revolution were not economists and did not have any economic plan of their own. In fact, they were glad to let the previous economic agents, with the exception of the above mentioned groups that were banned from their previous activities, continue their activities as before. The exodus of capital owners and some elite from the country would have been a major blow to the economic stability of the country, had it not been for the fact that there was only a relatively small production capacity and many goods were imported or produced by small entrepreneurs. The government had to reestablish the import channels and try to restore the previous standard of living, in spite of the sabotages that were surfacing as expected. To make matters worse, in November 1979, militant students took the personnel of the United States embassy as hostages. This in turn, was followed by United States retaliation and economic sanctions including freezing Iranian assets in the United States. The American embargo had a major impact on the Iranian economy and living standard given the United States was Iran's main source of imports, it housed most of Iran's assets abroad, and it was the main purchaser of Iranian oil. In order to overcome the blow and to avoid domestic unrest, the government had to take a very active role in all aspects of the economy, from imports to distribution (Abrahamian 2008, Amuzegar 1992, 1997, Daniel 2001, Nowshirvani 2010, Pollack 2005, Razi 1987).

To make matters worse, in September 1980, Iraq invaded Iran. Many Iranians were convinced that Saddam Hussein was encouraged by the United States to bring Iran down to its knees. The claim was given credence later when the United States and its close allies (such as Saudi Arabia, Persian Gulf Emirates, and other oil rich Arab countries) provided support to Iraq. Although it seems that all the assistance was due to "Arab solidarity," such solidarity was lacking in the case of the conflict between the Palestinians and Israel. The claim becomes less meaningful in light of the fact that in the case of the former an Arab country was the aggressor; in the case of the latter a non-Arab country was the aggressor.

The Iraq-Iran war engaged most of Iran's resources. Coupled with United States sanctions, the war took a toll on Iran's economy. Whatever structure was left after the revolution and the American embargo vanished as war consumed the country. Consequently, the Iranian government took control of many of the country's economic aspects. An Islamic government by nature is against a central economy. Individual property rights are very strong in Islamic tradition, and merchants are especially valued as the Arabs of the era before Islam and during the early days of Islam were primarily merchants, including the Prophet Mohammed.

After the onset of the Imposed War, the Iranian government began a rationing program. All consumer goods were subject to ration, which was conducted via coupons that were distributed based on the number of family members. Regardless, there were still long lines at distribution centers. Therefore, limitation of imports, lack of foreign currency, loss of oil production, and resultant loss of revenue led to the replacement of the price mechanism with coupons and waiting in lines (Abrahamian 2008, Amuzegar 1992, 1997, Daniel 2001, Nowshirvani 2010, Pollack 2005).

The economic relationship between "Islam" and economy has a dual nature. On one hand, the religion supports and encourages individual entrepreneurship, especially in the form of trade. While on the other hand, there has been a long history of ownership of property and business by Islamic institutions in the form of Waqf (Ashraf 1969, Sadeq 2002, Timur 2001). Waqf is an Arabic word, which means "religious endowment". A piece of property or an asset is Waqffed to a religious establishment. This means that the establishment is responsible for the upkeep of the property, or the asset, and in using it in its most productive way. The proceeds of Waqf are used for religious purposes, mostly to assist the poor. There were many reasons for the Waqf, such as religious belief, lack of children or relatives, protection from rulers, and good deeds. Many Iranian religious centers and shrines have vast amounts of Waqf. It is claimed that the Astan Guds Razavi, which is responsible for the upkeep of the Imam Reza's shrine in Mashhad, is the largest property owner in Iran. Therefore, private entrepreneurship is cherished and valued and it is normal for religious institutions to be owners of properties and assets. After the revolution, the government of the Islamic Republic viewed itself as the legitimate religious entity in the country and has not had any problem acting as an economic agent playing a major role in the country's economy. Economic sanctions and the Imposed War provided the need and the excuse as well. In addition, there are claims that some of the religious leaders in the government, and their families, are among the beneficiaries of the economic embargo and the need to centralize importation and distributions of the goods. In recent years, especially since Khatami's presidency the country has begun a widespread privatization effort by selling government-owned companies, factories, and corporations, some of which were in the hands of government prior to the revolution.

Ehteshami (1995: 84-85) reports that there were 41 families that owned major production and distribution facilities that were nationalized after the revolution. The majority of the listed firms are in consumer goods, mostly in the form of assembly. The factories that were not assembly-based include cement, glass, sugar, and textile production to name a few. There are two important related issues. First, none of these operations alone or together produced enough to meet domestic market needs; second, the necessary machinery for production was imported. The above statement is not meant to belittle the wealth or the power of any of these families. Quite the contrary, these families were very rich and powerful.

Ehteshami (1995: 85-86) reports that "… in 1978 the Lajevardi empire owned the Behshahr Industrial Group, which comprised 20 wholly-owned companies and 26 partnership ventures, in addition to having substantial stakes in the International Bank of Iran and Japan and in the Iranian Development and Investment Bank." In the late 1970s, the group had 12,000 employees and just one of its products (detergents) controlled about 30 percent of the domestic market. At the time, Iran was a country of 35 million people with low per capita income. Ehteshami (1995: 86) reports that after the collapse of the Pahlavi Dynasty "the court's holding company," the Pahlavi Foundation had to be taken care of. In response, the new government established the Foundation of the Deprived (Bonyad Mostazafan). Some argue that this foundation is used to finance the activities of the Hezbollah and other activities, which no one is quite aware of their details. However, "… this foundation was said to be ready, as early as 1983, to return some of the companies under its control to their former owners" (Ehteshami (1995: 86). Ehteshami does not elaborate who the previous owners were. The Pahlavi Foundation was owned by the Pahlavi family, and it is doubtful that the author is referring to them.

According to Ehteshami (1995: 86), the foundation and others similar to it have vast holdings. The important issue is the orientation of the regime. There is no evidence that the Islamic Republic government had or has any intention of participation in the production of goods and services (Ehteshami 1995: 85). In fact, within four years of its establishment, it was trying to privatize factories and other components of the production capacity of the nation. However, the pace of progress differed substantially under different presidents. The only thing that is certain is that properties of the royal family and others that fled the country were nationalized. There was also a provision to nationalize firms owing more than half of their assets to the banks (Ehteshami 1995: 86). states that the purpose of the latter was to rescue firms in financial trouble. It seems that only in the case of banks, and to a lesser extent insurance companies, did the government want to establish national control for the sake of control. In this case, control over the financial life of the nation. Especially in the early days of the revolution, this was important in order to curb capital drain from the country as well as to identify major capital owners and potentially corrupt individuals. Ehteshami (1995: 87) reports that $2-3 billion was taken out of the country, in the year leading to the revolution and over $15 billion in capital flight took place soon after the revolution.

The revolutionary components of the regime were in favor of nationalization of not only the factories and major production facilities, but they also supported confiscation of the assets of the "oppressors" on behalf of the oppressed. This view was very important, enough for the regime and supporters to address it in article 44 of the 1979 constitution. Article 44 states:

> Economic structure of the Islamic Republic of Iran is based on three sectors of public, co-operative, and private sectors with structured and appropriate planning. The Public sector consists of all large industries, principal industries, foreign trade, large mining, banking, insurance, energy, dams and large irrigation

systems, radio and television, postal, telegraph, and telephone, airline, shipping, roads and railroads and similar will be owned publicly and will be under government control. The private sector consists of the portion of agriculture, animal husbandry, industry, trade and services that complement economic activities of the government and the co-operatives. Ownership in these three sectors as far as they are in accordance with other articles of this chapter and do not exceed the legal boundaries of the Islamic laws and promote the growth and development of the country's economy and do not harm the society will be protected by the Islamic Republic laws. The law will determine and establish the extent and the conditions for each of the three sectors.

Even before the revolution, people in Iran considered the government to be responsible for the welfare of the people. This is a role that the Iranian government has been playing for a long time. However, Islam is a religion that supports and nurtures private ownership. The Islamic Republic has accelerated its pace of privatization. The tug-of-war between privatization and nationalization reflects the internal struggle of the still forming government. It reflects the give and take between revolutionaries and the clergy. This does not indicate by any means that the clergy is against government control of major industries and activities with negative or positive spillovers. The inherent problem of countries with Dutch Disease is their inability to compete with other countries due to overvalued currency. In the case of Iran, the Western embargo has taken part of the purchasing power of the foreign exchange by forcing the country to resort to the black market (in the case of weapons systems for national defense) or to buy its shortages from more expensive resources. Furthermore, embargos have increased the overall transaction cost by increasing what economists call the "sole of the shoe" problem, since it takes more effort (in the form of increased search, negotiations, etc.), transactions are more costly. Although over time the original revolutionaries have lost ground with regard to the issue of nationalization and the supporters of privatization seems to have the upper hand, the country's inability to compete in the global market necessitated some sort of government intervention. In the early days of the revolution, intervention was in the form of nationalization, but later it reverted back to subsidies and protections.

An indication of the government's orientation is the fact that, from 1979 to 1985, some 14,000 new factories and workshops employing over 30 people were established, mostly with government loans (Ehteshami 1995: 92). Ehteshami does not provide any indication of the annual rate of growth from these kinds of establishments. This potentially powerful entrepreneurial basis was nullified by rapid population growth and the Imposed War, exacerbating shortages and the need to increase imports. As one would expect, the Western embargo and Iran-Iraq war exacerbated economic problems associated with typical revolutionary years, resulting in negative growth through 1988 (Abrahamian 2008, Amuzegar 1992, 1997, Daniel 2001, Nowshirvani 2010, Pollack 2005).

The 1986-87 oil crises did not help the nation either. It is important to keep in mind that the majority of Iranian "industrial production" before and after the revolution depended on imported "intermediate goods" that could not be produced domestically. Obtaining intermediate goods was much more difficult and more expensive due to the Western embargo. Financial exigencies finally forced the government to borrow over $2 billion from abroad in 1989. This was in sharp contrast to the investments in foreign firms some 15 years earlier.

Rafsanjani's Presidency

During Rafsanjani's presidency, tax collections increased substantially. In this period, the conservative right wing gained ground. The group consisted of "The Society of Educators", "Qum seminary", "The Warrior Religious Society of Tehran", "Islamic Forums", "trades", "Islamic Economic Organization", "Guardian Council", "Hojjatieh Society", and "Confederated Council." This wing managed to curb the power of the radical wing of the government. The group under Rafsanjani's leadership managed to establish many new and first time innovations in the government of the Islamic Republic of Iran. Although the wing was not in favor of cultural or religious changes, nevertheless, it supported a free and open economy, which could be considered further towards the right than most European or United States examples.

Rafsanjani, in February 1989, stated that Iran will accept foreign loans provided they are not used for colonial ambitions. By appointing people to economic positions based on their ability to replicate other capitalist economies, both developed and underdeveloped, he demonstrated that he was a pragmatic man. MEED (June 14, 1991) reported Velayati, the Foreign Minister of Iran, as saying "economic considerations overshadow political priorities." During Rafsanjani's tenure the country could concentrate on economic growth and welfare. The crisis of revolution and the long war were in the past, although the Western embargo was alive and present. In 1991, the government began a program of denationalization, and promotion and encouragement of private investment. The stock market was reopened, and soon the trade volume surpassed the highest prerevolutionary level. By 1991, the volume of trade was more than 3.5 times the prerevolutionary level. During this period, almost everything was offered to private investors from mines to automobile assembly factories. A new export bank was established, and the door for foreign investment was opened. The former is evidence of government attitude. The Iranian government has always had an active role in promoting trade and investment. Active participation and aid from the government to promote growth is not an unusual thing in many developing countries. The policy had been successful in several countries especially in South Korea and Brazil. To encourage private and foreign investment, five free trade zones (FTZ) were created, enjoying most (if not all) the privileges and benefits of such FTZ worldwide. Finally, in 1993 Iran joined the General Agreement on Tariffs and Trade (GATT).

The economic fortune of Iran turned for the better during this period. It would be simplistic to attribute the rapid economic growth of over 12% in 1990, or the average growth rate of 8% for the first Five Year Economic Development and Expansion Plan, to one factor. During this period, the war ended, the price of oil increased, privatization was promoted, the population growth rate declined substantially, foreign direct investment began and grew rapidly, Iran joined GATT, The IMF funneled substantial sums of money, and the stock market was revitalized. This is just a part of what had changed from a few years ago. It is not clear which factor was the most important one, or played a more significant role. An objective and comprehensive study would be necessary to answer that question. It would be a safe statement to assert that all of these and other factors played some role in the change of economic performance from a declining economy to a growing economy.

Increased prosperity and ease of import resulted in increasing imports. Imports were financed by oil revenue as the primary and dominant source of foreign exchange. The increases in price of oil in 1989 and 1990 were substantial, and fueled the rush to import. Import increase was essential for the miracle growth since the majority of intermediate goods for assembly production in the country were imported. The price of oil increased some 50% in these two years. Starting with 1991, the price of oil began to decline once again. The drop was substantial from 1992 to 1993; it fell from $19.25 to $16.75. The decline was even more severe if it is adjusted for inflation. The drop forced the government to limit imports and thus lower economic growth. The result of increased imports, acceptance of foreign investments and loans, and the sudden reduction of oil prices was that Iran found itself unable to pay its notes on foreign loans.

In 1993, the Rial was devalued. Ordinarily, devaluation of currency would help industry especially the exporting industry. This was not and will not be the case for Iran. As mentioned, Iranian industries depend on foreign parts and intermediate goods. Devaluation makes those imports more expensive; therefore, resulting in decreased production. A decline in production translates into lower growth rates and hence lower incomes for the people. The government lowered the tariffs to counter the problem. Nevertheless, prices began rising and inflation was added to the list of the country's problems. The FTZs could not break the vicious circle of dependence on foreign intermediate goods nor could it produce goods that could compete in the global market. The problem, as pointed out earlier, stems from the impact of massive oil revenue that distorts the economy in a typical Dutch Disease way. Although the FTZs could not overcome the side effects of the disease, they had no problem becoming a major channel for imports. FTZs took advantage of all the benefits bestowed upon them; thereby, exacerbating the symptoms of the disease. Other contributing factors were Afghan refugees exceeding 4 million people (the largest in the world) exacerbated by natural disasters such as earthquakes and floods. Another major step was to increase and implement the tax laws as both a mean of obtaining revenue for the government and also as a mean of income distribution. Finally, the government began the process of abolishing rationing and subsidies, and to decrease the size and footprint of the government in the economy.

As a result of the fifth Majlis's election, there was a rift in the rightwing coalition. The religious and bazaar groups insisted on maintaining Islamic values and social justice, while the newly formed groups of "Development Brokers" ("Kargozaran Sazandegy", in Farsi) were promoting economic growth. This group consisted of a more leftist group of the rightwing coalition, and was supported by the elite and students. The former won the majority in Majlis forcing the latter to make a coalition with the left in the form of a "Hezbollah Assembly" (Party of God).

The economic relations between the United States and Iran have been through a series of stages beginning with the initial contacts where trade was primarily based upon American interests in commodities and Iranians who could afford luxury goods. Woven into this story is the game the great powers played with Iran as a prize in the global quest for influence. The moves of the great powers reflect the domestic political structure and its particular policy preferences. For the most part the great powers sought influence and access to commodities. The Cold War added communism to the equation and provided a basis for leaders like the Shah to increase Iran's military size and sophistication through large weapons purchases. The expenditure on what the common Iranian saw as extravagances that only enriched foreigners and other nations was not lost on the opposition who used such sentiments not only to overthrow the Shah but to create the Islamic Republic where such abuses would not take place. On the other hand the Untied States saw the Shah as the protector of the Persian Gulf and ignored the popular unrest his rule was causing. Both the United States and the Shah's government tended to dismiss any popular resentment toward the Shah as radical propaganda or communist agitation. The domestic political situation in both the United States and Iran was one where American interest in the oil supply were paramount and the arming of the Shah's military was seen as arming a friend. Furthermore, buying American weapons recycled the oil dollars and helped put Americans to work. The Shah and his supporters made small fortunes on various contracts with American and Western firms who helped modernize the Iranian economy. What the Shah failed to realize was that jobs and income was going to foreigners not Iranians and that this situation created massive resentment.

The roots of the revolution can be seen as not only political but economic as well. Economic stagnation for the vast majority of Iranians was not acceptable when wealth was seen as being siphoned off by foreign workers. American and western firms played into this perception by pressing for high wages and immunity from local laws. With this level of resentment and the subsequent revolution it follows that post-revolution relations would not be good between the United States and Iran. As history has demonstrated relations have indeed worsened and are at the level of a low level conflict. The continuation of the conflict is now part of the revolutionary narrative and for Americans Iran has been institutionalized as the proverbial "Bad Guy" To be sure, the truth lies somewhere in the middle between these two extremes. The fact, that trade may be a way to lessen hostilities has not been a policy option for either side as such a policy would be seen as a sign of weakness. Domestic politics in both countries seems to drive policy, rather than economic principles.

Bibliography

Abrahamian, E., 1970. Communism and Communalism in Iran: The Tudeh and the Firqah-I Dimukrat. International Journal of Middle East Studies, 1(4), 291-316.

Abrahamian, E., 1978. Factionalism in Iran: political groups in the 14th Parliament (1944-46). Middle Eastern Studies, 14(1), 22-55.

Abrahamian, E., 2001. The 1953 Coup in Iran. Science & Society, 65(2), 182-215.

Abrahamian, E., 2008. A History of Modern Iran, NYC: Cambridge University Press.

Alizadeh, Parviz, "Iran National Company", Encyclopedia Iranica Online, 2006, Available at www.iranicaonline.org

Amerie, S.M., 1925. The Three Major Commodities of Persia. The Annals of the American Academy of Political and Social Science, 122, 247-264.

Amuzegar, J., 1958. Point Four: Performance and Prospect. Political Science Quarterly, 73(4), 530-46.

Amuzegar, J., 1992. The Iranian Economy Before and After the Revolution. Middle East Journal, 46(3), 413-425.

Amuzaegar, J., 1997. Iran's Economy and the US Sanctions. Middle East Journal, 51(2), 185-199.

Ansari, A., 2001. The Myth of the White Revolution: Mohammad Reza Shah, 'Modernization' and the Consolidation of Power. Middle Eastern Studies, 37(3), 1-24.

Araghi, F.A., 1989. Land Reform Policies in Iran: Comment. American Journal of Agricultural Economics, 71(4), 1046- 1049.

Ashraf, A., 1969. Historical Obstacles to the Development of a Bourgeoisie in Iran. Iranian Studies, 2(2), 54-79.

Badakhshan, A., & F. Najmabadi, "Oil Industry ii. Iran's Oil and Gas Resources", Encyclopedia Iranica Online, 2004, Available at www.iranicaonline.org

Behrooz, M., 2001. Tudeh Factionalism and the 1953 Coup in Iran. International Journal of Middle Eastern Studies, 33, 363-382.

Bonakdarian, M., U.S.-Iranian Relations, 1911-1951. In The United States and the Middle East: Diplomatic and Economic Relations in Perspective. Yale Council on Middle East Studies, pp. 9-25.

Carey, J.P. & Carey, A.G., 1976. Iranian Agriculture and Its Development: 1952-1973. International Journal of Middle East Studies, 7(3), 359-382.

Carnoy, Martin, and Thias, Hans H. "Draft Report of Second Tunisia Education Research Project RP0248." Mimeographed. Washington, D.C.: I.B.R.D., 1974.

Clapp, G.R., 1957. Iran: A TVA for the Khuzestan Region. Middle East Journal, 11(1), 1-11.

Connell, J., 1974. Economic Change in an Iranian Village. Middle East Journal, 28(3), 309-314.

Dadkhah, K., 2001. The Iranian Economy during the Second World War: The Devaluation Controversy. Middle Eastern Studies, 37(2), 181-198.

Daftary, F., "Barnama-Rizi", Encyclopedia Iranica Online, 1988, Available at www.iranicaonline.org

Daniel, E., 2001. The history of Iran, Westport, CT.: Greenwood Press.

Davis, S., 2006. "A Projected New Trusteeship"? American Internationalism, British Imperialism, and the Reconstruction of Iran, 1938-1947. Diplomacy and Statecraft, 17, 31-72.

Embry, J., 2003. Point Four, Utah State University Technicians, and Rural Development in Iran, 1950-64. Rural History, 14(1), 99-113.

Floor, Willem, "Steel Industry in Iran", Encyclopedia Iranica Online, 2005, Available at www.iranicaonline.org

Foran, J., 1989. The concept of dependent development as a key to the political economy of Qajar Iran (1800-1925). Iranian Studies, 22(2), 5-56.

Freivalds, J., 1972. Farm Corporations in Iran: An Alternative to Traditional Agriculture. Middle East Journal, 26(2), 185-193.

Galpern, S.G., 2002. Britain, Middles East Oil, and the Struggle to Save Sterling, 1944-1971. Middle East.

Gavin, F.J., 1999. Politics, Power, and U.S. Policy in Iran, 1950-1953. Journal of Cold War Studies, 1(1), 56-89

Ghaneabassiri, K., 2002. U.S. Foreign Policy and Persia, 1856-1921. Iranian Studies, 35(1/3), 145-175.

Greaves, R.L., 1968. Some Aspects of the Anglo-Russian Convention and itws Workings in Persia, 1907-1914--II. Bulletin of the School of Oriental and African Studies, University of London, 31(2), 290-308.

Harris, F.S., 1953. The Beginnings of Point IV Work in Iran. Middle East Journal, 7(2), 222-228.

Hetherington, N.S., 1982. Industrialization and Revolution in Iran: Forced Progress or Unmet Expectation? Middle East Journal, 36(3), 362-373.

Karshenas, M., & H. Hakimian "Industrialization ii. The Mohammad Reza Shah Period, 1953-79", Encyclopedia Iranica Online, 2004, Available at www.iranicaonline.org

Kazemi, F., "Anglo-Persian Oil Company", Encyclopedia Iranica Online, 1985, Available at www.iranicaonline.org

Kristjanson, B.H., 1960. The Agrarian-Based Development of Iran. Land Economics, 36(1), 1-13.

Ladjevardi, H., 1983. The Origins of U.S. Support for an Autocratic Iran. International Journal of Middle Eastern Studies, 15(2), 225-239.

Lewis, W. Arthur (1954). "Economic Development with Unlimited Supplies of Labor," Manchester School of Economic and Social Studies, Vol. 22, pp. 139-91.

Looney, R., 1988. The role of military expenditures in pre-revolutionary Iran's economic decline. Iranian Studies, 21(3), 52-83.

Mahdavy, H., 1965. The Coming Crisis in Iran. Foreign Affairs, 44(1), 134-146.

Marsh, S., 1998. The Special Relationship and the Anglo-Iranian Oil Crisis, 1950-4. Review of International Studies, 24(4), 529-544.

Millspaugh, A.C., 1933. The PErsian-British Oil Dispute. Foreign Affairs, 11(3), 521-525.

Nowshirvani, Vahid, "Commerce vii. In the Pahlavi and post-Pahlavi periods", Encyclopedia Iranica Online, 2010, Available at www.iranicaonline.org

Paine, C. & Schoenberger, E., 1975. Iranian Nationalism and the Great Powers: 1872-1954. MERIP Reports, (37), 3-28.

Pesaran, M. Hashem, "Economy ix. In the Pahlavi Period", Encyclopedia Iranica Online, 1997, Available at www.iranicaonline.org

Pirouz, K., 2001. Iran's Oil Nationalization : Musaddiq at the United Nations and His Negotiations with George McGhee. Comparative Studies of South Asia, Africa, and the Middle East, 21(1/2), 110-117.

Pollack, K.M., 2004. The Persian Puzzle: The Conflict Between Iran and America, New York: Random House.

Pryor, L.M., 1978. Arms and the Shah. Foreign Policy, (31), 56-71.

Rahnema, S., 1990. Multinationals and Iranian Industry: 1957-1979. Journal of Developing Areas, 24(3), 293-310.

Ramazani, R.K., 1974. Iran's 'White Revolution': A Study in Political Development. International Journal of Middle East Studies, 5(2), 124-139.

Razi, G.H., 1987. The Nexus of Legitimacy and Performance: The Lessons of the Iranian Revolution. Comparative Politics, 19(4), 453.

Ricks, T.M., 1979. U.s. Military Missions to Iran, 1943-1978: The Political Economy of Military Assistance. Iranian Studies, 12(3), 163-193.

Ronfeldt, D., 1978. Superclients and Superpowers: Cuba:Soviet Union/ Iran: United States, Santa Monica, CA.

Rubin, M.A., 1995. Stumbling through the " Open Door ": The U.S. in Persia and the Standard-Sinclair Oil Dispute, 1920-1925. Iranian Studies, 28(3), 203-229.

Sadeq, A.M., 2002. Waqf, perpetual charity and poverty alleviation. International Journal of Social Economics, 29(1-2), 135-151.

Seitz, J.L., 1980. The Failure of U.S. Technical Assistance in Public Administration: The Iranian Case. Public Administration Review, 40(5), 407-413.

Shoamanesh, S.S., 2009. Iran's George Washington: Remembering and Preserving the Legacy of 1953. MIT International Review.

Summitt, A.R., 2004. For a White Revolution: John F. Kennedy and the Shah of Iran. Middle East Journal, 58(4), 560-575.

Timur, K., 2001. The Provision of Public Goods under Islamic Law: Origins, Impact, and Limitations of the Waqf System. Law & Society Review, 35(4), 841-898.

Volodarsky, M., 1983. Persia and the Great Powers, 1856-1869. Middle Eastern Studies, 19(1), 75-92.

Young, T.C., 1962. Iran in Continuing Crisis. Foreign Affairs, 40(2), 275-292.

Zahrani, M.T., 2002. The Coup that Changed the Middle East: Mossadeq v. the CIA in Retrospect. World Policy Journal, 93-100.

Chapter 6

The 1979 Revolution and the Beginning of the Conflict with the United States

The Revolution of 1979 brought the low-level conflict between the United States and Iran, which was brewing for over 30 years, to the surface. The selectorate that kept the Shah in power lost its privileged position in the political system only to be replaced by radicals who sought to change society in all aspects. The change in leadership empowered those who had little power under the Shah and also changed the foreign policy of the new Islamic Republic.

By all accounts the Iranian revolution of 1979 was unpredicted, and in most cases unexpected. There seemed to be very little reason for a revolution and even less evidence for it. What was even more unexpected was the fact that it would be a religious-based one. In fact, the Islamic Republic of Iran (IRI) is one of the few countries in the world that is actually based on a religion and the only Shi'a country in the world. The event is so rare that only one other such government in the world has ever existed: the Safavid Dynasty (1502-1722) in Iran, which of course was much larger than the present-day Iran and stretched from eastern parts of Afghanistan to western parts of Iraq and deep into the center of Russia (Duby 1987). Consequently, their new republic did not have any natural allies. In fact no one knew what an Islamic Republic should do and what it should or should not do differently than other countries. Initially, it was obvious that the IRI must be based on Islam. Consequently, all activities and decisions had to be examined through an Islamic lens to determine if they conform to the teachings of the religion and that they did not violate any Islamic laws and principles (Abrahamian 2008, Green *et al.*2009, Rakel 2009, Schahgaldian 1989, Thaler *et al.*2001)

Another expected and obvious outcome was the need to change the relationship with the United States. The role of the later in the Codetta of 1953 had resulted in great power and influence in Iran. In fact, Iranians viewed the Shah as a puppet of the United States and opposition documents and statements, most notably those of Khomeini, referred to him as the "American Shah" (Cooper & Tefler 2006, Jordet nd, Meskill 1995, Quosh 2007). The 1964 agreement between the two governments granting American citizens, regardless of their function in Iran or purpose for being in Iran, capitulation privileges ended any doubts that one could have had about the legitimacy of the Shah's regime (Meskill 1995). Anytime that a country receives a unilateral privilege it is an indication of dominance. This is very different than agreements that extend similar, if not equal, privileges for both parties (Renton 1933).

Both the United States and Iran knew that the relationship between the two countries would change. The only thing uncertain was the extent of the change. The Islamic Republic of Iran ended all the privileges that were given to the United States; however, it let some contractual agreements stand. In fact, it demanded the United States deliver the weapons that Iran paid for and were awaiting production (Abrahamian 2008, Meskill 1995).

Iran not only curtailed its relationship with the United States but was also completely cut-off from the rest of the world either willingly or unwillingly. Iran maintained its diplomatic relationship with almost all the countries with which it had relationships before the revolution. In fact, at first, the United States Embassy continued its normal routine with over 1,000 staff and diplomatic personnel. In order to form and shape its "Islamic" identity, Iran felt compelled to pave its own path. The slogan of "neither West, nor East," reflected the orientation of the leaders. The slogan was designed to not only reject Western ideals and cultural values, primarily the influence of the United States, but also to assure that the Soviet Union (and for that matter any other socialist/communist country) could not establish any power base in Iran (Abrahamian 2008). Leaders such as Khomeini still remembered the days of Soviet influence through the "Tudeh" communist party and the declaration of independence by Azerbaijan and Kurdistan (Behooz 2001, Chaqueri 1999). The memories of Tsarist Russia's meddling in Iranian's affairs and virtual colonization of the country by Russia and Great Britain were still alive ("Russia and Persia" 1912, Tapp 1955). Consequently, Iran became isolated because it did not have any natural allies. Contrary to the common belief, Muslim countries were not and still are not allies of Iran. There are several reasons for this, such as dependence or close ties to the United States and a different interpretation of Islam. In fact, since at the time of Iran's revolution no Muslim country had a democratically elected government. Many Arab countries, were concerned that an Iranian-type revolution might end their regimes.

In 1979 the only countries with sizable Muslim populations conducting elections were Turkey, Indonesia, Malaysia, Brunei, and Albania. Turkey is the only country close enough to be affected. Although at the time Turkey did not have any reason to be concerned about support for an Islamic government, the situation since then has changed and the possibility is no longer as remote as it seemed in 1979. Arab countries, especially those nearby, such as Kuwait, Bahrain, and the Emirates can and actually are influenced by the events in neighboring larger countries such as Iran and Saudi Arabia, regardless of their respective regime types. The Islamic Republic with its promise of democratic elections, Islamic values, and by virtue of overthrowing a dictator rattled the palaces of the leaders of countries from Saudi Arabia to Pakistan. It also raised the hopes of millions of the possibility to be independent from the tyranny of their dictators and subservience to the West. Some people even envisioned the possibility of the return of Islam's glorious days when the Muslim Caliphate stretched from India to Spain. Some of the external reasons for Iran's isolation stemmed from this possibility. Obviously, Iran was promoting such thoughts and thus was considered a natural leader for the cause.

The attack by Iraq did not help the isolation of Iran either. A discussion of the reasons and motives for the attack are beyond the scope of the present book. The war helped to increase the isolation and also helped to end it. The war proved that when one side is receiving weapons and other aid from numerous sources, no country can stand alone. Consequently, some of the ideological standards had to be bent for sake of pragmatism to meet the needs of the country and the citizens to ensure their survival. Iran first turned to communist countries, especially to Vietnam, to get parts for planes and weapons which they had seized after defeating the Americans in 1975. Since supplies were not sufficient Iran was forced to deal with the United States to obtain weapons. This was in spite of the ongoing harsh stance by both countries against each other and with full knowledge of Ayatollah Khomeini himself (Iran-Iraq War *nd*).

The Iranian revolution was a blow to the regional hegemony of the United States. The decade of the 1970s was not very good for the United States. In 1973 OPEC managed to flex its muscle against the West, particularly the United States. Regimes such as Saudi Arabia and Iran, the two most powerful and largest countries in the organization, finally turned on the United States and took a harsh and rigid stance on oil prices (Danielsen 2010, Wonnacott 2010). Then, in 1975, the Vietnam War ended with the capture of Saigon by communist troops (Duiker & Turley 2010, Spector 2010). In 1979 the biggest and most powerful ally of the United States in the Middle East, Iran, declared its independence and ended 185 years of Western dominance in Iran. In the same year the Soviet Union invaded Afghanistan. However, things began to improve for the United States in the 1980s when in September 1980 Iraq attacked Iran and the United States kept Iran focused on a border war by helping Saddam. The end of the decade was really good for the United States. In 1988 the Soviets, admitting defeat, began withdrawing their troops from Afghanistan (Dewdney 2010). In 1989 the Berlin Wall was demolished (Erb & Reuter 2010). Iraq's invasion of Kuwait in 1990 sent Arab countries back to the United States seeking help and assistance. Soon, in 1991, the Soviet Union imploded and all the communist countries of Eastern Europe followed suit immediately. Consequently, numerous new countries in Europe and Asia were formed and almost all became United States allies and established close ties with the West.

Iran US Relations

The Iranian revolution changed the geopolitical map of the Middle East. Every regime in the region became nervous. The fact that Iranians were advocating similar uprisings in other countries added to the fear. The United States lost a very dependable and strong ally together with all the listening devices, other spy tools, and posts that enabled it to monitor part of the Soviet Union. After a successful referendum in support of an Islamic Republic and establishment of a constitutionally elected government, the remaining issue seemed to be demand for the Shah's return to justice, which the country and Ayatollah Khomeini would not let go. The only country that was willing to admit the Shah was Egypt. President Anwar Al-Sadat of Egypt actually gave a warm welcome to the Shah. Consequently, Iran severed its diplomatic ties with Egypt and dismissed Egyptian diplomats. The Shah's stay in Egypt was temporary because of Anwar Al-Sadat's weak government and lack of support among Egyptians. In fact none of the United States allies in the Middle East wanted to have anything to do with the Shah. Consequently he was moved to Morocco, Bahamas, and Mexico (Abrahamian 2008, Daniel 2001).

The United States had a major dilemma. There were many non-democratic countries around the world that were watching how the United States was dealing with the Shah. The fact that the United States did not keep the Shah in power was a major eye opener for other dictators that depended on the United States for survival. If the United States refused to admit the Shah into the country it would have further alienated the remaining dictators. On the other hand, if the US allowed the Shah entry into the country it would definitely lose any hope of reconciliation with Iran. Finally, President Carter decided to allow the Shah entrance into the country for cancer treatment. This seemed a reasonable humanitarian gesture and Iran had been informed that the stay would be temporary. However, Iran refused to believe the medical necessity and insisted that Iranian physicians examine him. They even promised to provide necessary medical attention as needed provided that he was surrendered to Iran to stand trial for crimes against humanity (Abrahamian 2008, Daniel 2001). Obviously, the United States could not have backed off and the Iranians were as relentless.

The Shah entered the United States on October 22, 1979. On November 4, 1979, less than 10 months since the return of Ayatollah Khomeini to Iran, some militant Muslim students stormed the United States Embassy in Tehran during a demonstration that was conducted in front of the embassy. The Hostage Crisis, as it became known in the United States, ended the official diplomatic contact between the two countries (Abrahamian 2008, Daniel 2001, Hamilton *et al.* 2001b, Torbat 2005). The United States refused to extradite the Shah and the "students" refused to release the hostages. Although the Shah left the United States on December 15, 1979, Iran refused to release the embassy personnel. There is no doubt that the Iranian government and Ayatollah Khomeini supported the occupation of the embassy and were involved in negotiations with the government of the United States (Daniel 2001, Roberts 1996).

Taking US embassy personnel hostage, and the inability of the United States to free them, was a symbol of the power of independence, which Iran exploited in order to win the hearts of the masses in many Third World countries (especially Muslim countries of the Middle East). From the very beginning Ayatollah Khomeini painted the Islamic Republic of Iran as the hope of what he called "oppressed people" inside and outside of Iran. The symbolic representation of the oppressor was the United States and President Carter as the sitting president was personification of the symbol. His "humanitarian" act of permitting the Shah to get medical attention in the United States was taken as an act of tyranny, ignorant of the feelings of oppressed Iranians. It is unlikely that the original justification for entering the embassy was based on the above discussion. Revenge would have been more justifiable. Even the idea of exchanging the hostages for custody of the Shah could have also been a motive. Of course the Shah's departure from the United States was considered by Iranians as a convenient excuse for the United States not to extradite him to Iran.

As a result of all these matters hostage negotiations were difficult. At times the Iranians played hardball and at other times the United States reciprocated the favor. The relationship between Iran and the United States got even worse by the occupation of the US embassy in Tehran. The Shah's death in July of 1980 made an exchange impossible. The ordeal had been going on for too long and it was becoming a liability for Iran (Daniel 2001, Torbat 2005). Both countries were trying to end the stalemate. In fact an agreement had been brokered and on September 22, 1980 the hostages were in an airplane at Tehran's airport when the airport was attacked.

At first the Iranians thought the United States had attacked again. Soon it became evident that it was Iraq that had invaded Iran and Iraqi troops were advancing rapidly, deep into Khuzestan in South-West Iran (Cordesman 2003, Daniel 2001). Immediately, Iran blamed the United States, and in turn, the US categorically denied any connections. However, the evidence shows that the United States was providing secrets and information about Iran's military capabilities, shortcomings, soft spots, and other logistic information to Saudi Arabia and Israel, which in turn were shared with Iraq (Hamilton et al.2001b, Quosh 2007). Consequently, the hostages were returned to hiding as Iran began organizing its defense against the invasion. This was the last straw for the Iranians and they never forgave President Carter to the point that even after President Carter was defeated by then Governor Reagan, Iran did not release the hostages until 4 minutes after President Reagan's inauguration into office (Abrahamian 2008, Daniel 2001, Pollack 2007). Such overtures were based on the realities of international affairs; as such, the two countries did not soften their stance against each other. The United States did everything in its power to overthrow the Iranian government, including an economic and trade embargo; freezing of Iranian assets; and aiding Iraq in the war against Iran by providing weapons, including weapons of mass destruction. The United States coordinated activities and policies with allies and friendly governments to make them appear legitimate.

The US was further discredited later when shredded documents of the embassy were pasted together and published in 54 volumes in 1982. If there were any doubts about the role of the United States in Iran, these documents ended it. The documents helped the government maintain a strong anti-American stance. Although by this time the country was engaged in war with Iraq, the stance against the United States and Israel provided a way of sustaining revolutionary attitude (Daniel 2001, Hamilton *et al*.2001b, Smith 2007, Thaler *et al*.2010, Torbat 2005, Wehrey *et al.* 2009b). At least in official statements the Iranian government stated that the restored documents proved the United States was meddling in Iranian affairs and thus the occupation of the embassy and severance of ties with the United States were justified (Daniel 2001, Thaler 2010). The United States, on the other hand, pointed out that the embassy of a country is considered part of the sovereign country and that it was up to the Iranian government to assure the security of the embassy and the safety of the American diplomats. In short, the hostage crisis was another reason for worsening of the relationships between the two countries.

The foreign policy of the Islamic Republic of Iran during its first decade was based on its war with Iraq; campaign against the United States; and the development and fine-tuning of the concept of an Islamic Republic. Such efforts greatly influenced the country's development throughout the decade (Abrahamian 2008, Daniel 2001, Roberts 1996, Thaler *et al.*2010). Ayatollah Khomeini was the undisputed leader of the country and personification of the revolution. Everything in this period was based on his interpretation and teaching of the religion and all other aspects of life, including foreign policy. Although Iran severed diplomatic ties with few countries, the pressure from the United States and its allies (who were also concerned about Iran's stance on freedom, revolution, and Islam) resulted in virtual diplomatic isolation.

This era signaled the beginning of the low-intensity fourth generation war between the United States and the Islamic Republic. The Iranian selectorate had changed. Instead of being willing to help the United States in its mission to secure the Persian Gulf it was hostile towards the US and supported countries with anti-west sentiments.

On the other hand, trade relations were never broken. Iran continued exporting its oil and importing almost all of the same goods as before the 1979 Revolution. Even imports from United States existed, which was listed in the statistics books under "other countries in America," which referred to the continent. Sometimes economic laws are like the gravity law of physics; it does not matter what ideology one follows, it is subject to the same economic laws. Contrary to popular belief, the Islamic Republic of Iran is neither rigid nor uncompromising. Iran began purchasing American made weapons from Communist Vietnam and had strong ties to China and the Soviet Union. Even under Ayatollah Khomeini's watch Iran purchased weapons from the United States directly. The fact that the United States entered into the contract demonstrated that the United States was as calculating in its foreign affairs and understood that there are times when what a country says is not necessarily what a country does (Daniel 2001, Hamilton *et al.*2001b, Hicks 1996, Kinsella 1994, Mistry 2003, Quosh 2007, Torbat 2005). Iran, as adamant in its righteousness, would not miss the opportunity to assist groups and organizations fighting against the United States or its close allies, especially Israel. Shi'a groups in Palestine and Lebanon were obvious beneficiaries. Opposition groups inside and outside Saudi Arabia benefited as well. It is worthy of mention that Saudi Arabia was one of a handful countries without diplomatic relations with Iran. Saudi Arabia was more fearful of Iran than the United States. The Iranian Revolution or a similar uprising was, and still is to some extent, a major threat to regimes such as the one in Saudi Arabia or Jordan. During the decade of the 1980s Iran was primarily involved with the war and consolidation of its grip on domestic power. Early in the decade the government managed to neutralize or isolate all domestic opposition (Abrahamian 2008, Daniel 2001, Roberts 1996, Wehrey *et al.*2009b). This does not mean that the country had completely been consumed with the war. On the contrary, Iran was doing everything it could to undermine the United States directly or indirectly through its allies, especially Saudi Arabia and Israel. In 1982, when Israel invaded Lebanon, Iranians were quick to help the Shi'a to make sure that the massacres of Tall-al-Za'tar in 1976 would not be repeated. Iran sent 1,000 revolutionary guards to South Lebanon to train Shi'a groups. This was the beginning of the Hezbollah (Hajjar 2002, Smith 2007, Wehrey *et al.*2009). The United States accused Iran of arming Hezbollah, which Iran has denied, while admitting to providing training and financial assistance. Iran has also been accused of enticing violence and agitation in Saudi Arabia. As a result of demonstrations by Iranians and natives during the Hajj, the relationship between Iran and Saudi Arabia was tense. One such event, in August of 1987, resulted in a strong reaction by Saudi forces leading to over 600 deaths (Cordesman & Al-Rodhan 2006, Hajjar 2002, Wehrey *et al.*2009b).

Regaining the lost territories from Iraq helped the government and people of Iran and provided a glimpse of hope. However, the war was dragged on for too long. Iran's ability to purchase weapons from the black market, Vietnam, and finally from the United States helped her to stop Iraq's advance and regain the lost territories. However, it was not sufficient to defeat Iraq. Since it was not in the interest of the United State for Iraq to win the war outright and have access to virtually all Iranian oil and since she could not see Iran as a victor the only option was to make sure that neither would win nor lose. The fact that the United States reversed its stance and instead of helping Iraq, as right before the war and in its early days, and sell weapons to Iran while continuing help to Iraq, indicates that the United States benefited from the continuation of the war, which weakened both countries (Hicks 1996, Quosh 2007). Without doubt, Iran's preoccupation with the war limited its ability to help sympathetic groups and organizations. The reality and the consequence of a stalemate war forced both Iraq and Iran to seek an alternative and finally sign a ceasefire agreement. Once again, Ayatollah Khomeini had to set his hard-line public stance aside and to agree with a ceasefire. The pragmatism shown by Khomeini reflects his consolidation of power over the selectorate and also signals the pragmatism of the ruling elites. Knowing how a continued war would drain the nation's resources further and that the population would become war weary they made the decision to terminate the conflict with Iraq, however Iran did not terminate its low-level conflict with the US. Indeed, without Iraq to worry about Iranian leaders could pursue the conflict with the United States unimpeded.

On April 14, 1988 the USS Samuel B. Roberts frigate hit a mine in the Persian Gulf, which was believed to be dropped by Iran to stop oil tanker traffic as part of its war effort with Iraq (Hamilton *et al.*2001b). Four days later, in the largest US naval operation since the World War II in retaliation, the United States warships opened fire on an Iranian frigate sinking it and then shelled two Iranian oil platforms. On July 3 of the same year the United States Navy shot down Iran Air Flight 655 claiming it was directly moving towards an American warship and was ignoring communication attempts. Later on, the United States called the incident an error and paid retribution to the families of the victims (Cordesman & Al-Rodhan 2006, Hamilton *et al.*2001b). The relationship between the two countries became tenser than ever.

In 1989 Ayatollah Rafsanjani was elected as President. He initiated some contacts with other countries and began reestablishing ties with countries both in the West and the East as well as underdeveloped countries. The mission of his government was to repair war damages, end shortages of virtually everything, starting with food, and to improve Iran's stance in the international community both economically and politically (Abrahamian 2008, Green *et al.*2009, Thaler *et al.*2010). As Iranians showed more pragmatism during Rafsanjani's presidency, the United States began a more hegemonic stance thanks to the defeat of the Soviet Union in Afghanistan and then its collapse two years later. The only major rival of the United States suddenly vanished, and newly formed Eastern European and Central Asian countries rushed to out-pace each other in making close ties with the United States. The European Union was an amalgamation of diverse countries, mostly disjoint, with power struggles between France and Germany for leadership of in the European Union. The only other remaining substantial power, namely China, was never powerful enough to pose a challenge to the United States or the Soviet Union for that matter. Obviously the defeat of Soviet Union troops in Afghanistan (with not-so overt help and supply from the United States) was welcome news to the Americans. A more pragmatic and less antagonistic new government in Iran was also beneficial. This does not mean that everything was fine between Iran and the United States. Quite the contrary, the two countries were openly criticizing each other and doing whatever possible to undermine each other and to gain the upper hand.

President Rafsanjani had shown his pragmatic approach to problems by recommending a ceasefire with Iraq. He also played a crucial role in choosing Ayatollah Khamenei as the Supreme Leader (Abrahamian 2008, Green *et al.*2009, Hamilton et al.2001b, Thaler et al.2010). There were high hopes for change in the relationship between Iran and the United States since Khomeini had passed away. Khamenei was believed to be less radical than his predecessor and now Ayatollah Rafsanjani was the second person in command. In the same year the United States had a new president. Although President Reagan's Vice President won the election, he was considered less hardliner than his predecessor. The new presidents began under difficult conditions. Iran was struggling with its economy after the prolonged war but had gained more power and prestige in the region. It also had more room to help sympathetic groups in the region. By then Hezbollah had become a formidable force in Lebanon especially with the weakening of the Palestinian Liberation Front (PLO) in the region in general (and in Lebanon specifically) and the uprising of the Intifada. The Iranian government established a support base and garnered influence in the region. Many hostages were taken in Lebanon, believed mostly by Hezbollah or other Shi'a factions. The United States and some European countries had claimed that there was evidence of Iranian support, if not outright involvement or order (Byman 2003, Cordesman & Al-Rodhan 2006, Hajjar 2002, Hamilton *et al.*2001b, Smith 2007, Thaler et al.2010, Torbat 2005, Wehrey et al.2009b).

Another factor that influenced the affairs of this era was the fact that the Iran-Contra scandal was still recent and took place during the Republican Party leadership. So, the new president (a republican) could not directly negotiate with, nor offer money to, Iran. Nevertheless, the United States had come to the conclusion that the key to freeing hostages in Lebanon was in Tehran. It could be also that in this political tango it was the turn of the United State to take the next step. In his inaugural speech, President George H. Bush took a conciliatory tone and sent a signal to Iran. He stated his concern for the missing Americans and those who were held hostage in Lebanon. He also stated that "assistance can be shown, and welcome and long remembered." He also made a promise: "goodwill begets goodwill." The President followed his message with a call to the United Nations Secretary Javier Perez de Cuellar, promising that if the Iranians could see to it that the United States hostages were freed from Lebanon, then United States would make concessions to Iran as long as there were no direct negotiations. The United Nations negotiator Giandomenico Picco visited President Hashami and gave him the message which was received well. The President promised that if things would get better for Lebanon and Palestinians, then Iran would do anything it could to free the hostages for humanitarian reasons (Abrahamian 2008, Hamilton *et al.*2001a, 2001b, Smith 2007, Thaler *et al.*2010, Torbat 2005, Wehrey *et al.*2009b).

In August 1990 events changed dramatically when Iraq invaded Kuwait and the United States began pouring troops and military equipment into the Middle East in earnest, making Iran very nervous with its presence. Iran, which had just ended a prolonged war with Iraq without being able to win the war, felt weak and was in no position to risk an attack by the only remaining super power. Iran increased its pressure and in the end all hostages were freed. Although it took three years for Iran to secure the release of the hostages, they felt that they had delivered their end of the bargain and were waiting for reciprocating favors from the United States in form of concessions as promised. Indeed, they contacted the United Nations Secretary and asked him to remind the US of the promise it had made (Abrahamian 2008, Hajjar 2002, Hamilton et al.2001a, Smith 2007, Thaler et al.2010, Torbat 2005, Wehrey et al.2009b).

The State Department was in favor of reciprocating the favor by removing some sanctions against Iran. However, the White House advisors were against it on the grounds that Iran increased its support for militant groups in the Middle East and elsewhere. It seems that the government of Iran was trying to get one on the United States. On one hand it pressured Hezbollah and other militant groups to release the hostages, and on the other hand it helped them with other activities and financial assistance. If the United States lifted sanctions in response to the hostage release then Iran would have been the clear winner of this gambit. This would have increased Iran's influence among the militant groups by increasing its support. Nevertheless, dictating the course of action they need to take (Abrahamian 2008, Crane et al.2008, Hamilton et al 2001a, Thaler et al.2010, Torbat 2005). This latter possibility was heavy on the minds of Iranian officials and citizens alike.

Amassing hundreds of thousands of troops in the Persian Gulf; having military agreements with Bahrain and United Arab Emirates; and establishing military bases in Saudi Arabia were sources of concern and the reason for Iranian pressure to finally release Western hostages in Lebanon. The United States' agreements with recently formed countries to the north such as Azerbaijan, Turkmenistan, and Uzbekistan were added concerns. It is possible that in order for Iran to secure the release of the hostages, it had to make concessions to the militant groups. After all, Iran does not command any of the militant groups. Its power of persuasion is limited even when it comes to Hezbollah, which was established with Iran's assistance and training (Byman 2003, Hajjar 2002, Smith 2007, Thaler et al.2010, Wehrey et al.2009b).

While the United States government was wrangling with its internal politics over how to respond to Iran's gesture of goodwill a bomb destroyed the Israeli Embassy in Buenos Aires, Argentina. The Islamic Jihad Organization with ties to Hezbollah and Iran claimed the responsibility (Byman 2003, Cordesman & Al-Rodhan 2006, Hajjar 2002, Smith 2007, Torbat 2005, Wehrey et al.2009b). This event split the American selectorate by strengthening the hand of those groups and politicians who sought a harder line with Iran and also convinced the undecided that Iran was supporting extremist movements. The position of hardliners in the United States government was that Iran should not be rewarded while it was increasing its support for militant groups that were actively engaged in opposing the United States or Israel and attacking the interests of these two countries and their allies. The refusal of the United States to keep its end of the agreement put President Rafsanjani on a slippery slope and weakened the position of moderates within Iranian governmental factions (Abrahamian 2008, Crane et al.2008, Green et al.2009, Hamilton et al.2001b, Torbat 2005). Either because of this weakening or because of President Rafsanjani's personal feeling of betrayal, Iran did not provide any more assistance or gestures of goodwill towards the United States. No real negotiation or serious contacts were made between the two countries as long as President Rafsanjani was in charge, which extended five more years and overlapped with President Clinton's first term in the office.

The set of events between President George H. Bush's inaugural speech and the end of the second term of President Rafsanjani demonstrates the delicate and cautious moves and counter-moves of both parties. It highlights the distrust between the two countries and their respective leaders. It also reveals the fact that each country has to deal with its internal and external limitations and opportunities while trying to deal with each other. During this time the United States became the only remaining super power in the world. Many strategic worries of the United

States evaporated when the Soviet Union crumbled. The US felt it could pursue a Western agenda of liberal democratic capitalism. On a smaller scale it seems that Iran felt is should be playing a major regional role (Crane *et al*.2008, Green et al.2009, Thaler *et al*.2010, Wehrey et al.2009b). Since these two objectives are mutually exclusive, and in fact clash due to differences in philosophy and perspective, the two countries find themselves facing each other in almost every issue. No sooner than one party does something in pursuit of its goals and objectives then the other one realizes that it threatens its position or endangers its interests. The era that should have been the time of undisputed American hegemony has been marred by Iran, Al Qaeda, and the Taliban.

The Taliban, which was formed with aid from Pakistan after the Soviet Union defeat in Afghanistan, had turned its effort into building a strict Islamic country, which opposed Western values and opposed the United States among other nations (Ghufran 2001, Khalilzad 1995, Magnus 1997, 1998, Rubin 1997). The creation of the Taliban was the result of foreign powers' desire to have a say in the internal affairs of Afghanistan and to secure their own interests. The opportunity arose from the fact that the Mujahidin, which were led by Ahmad Shah Massoud and others, failed to secure the country. In-fighting continued giving Pakistan an opportunity to elevate its supporters in Southern part of Afghanistan and provide them with funding and equipment. The United States was not that unhappy with the Taliban because Ahmad Shah Massoud had gained respect and power and was gaining momentum to dominate the political scene. Since Iran was the main supporter of Ahmad Shah Massoud, it created anxiety for the United States. From the very beginning the relations between Iran and the Taliban was tense. After all, one purpose of creating the Taliban was to reduce or eliminate Iran's influence over the Mujahidin, and to curb the influence of Shi'a factions such as Ahmed Shah Massoud. As early as September 1995, Iran was issuing warnings to the Taliban not to cross Iranian boarders. In September of 1996 when the Taliban captured Kabul and large territories south of it, Iran in unison with Russia, India, and a host of Central Asian countries condemned the event. In October 1996, Iran helped Ismail Khan in Maimana (on the North West of Afghanistan) against the Taliban while continuing its support for Ahmed Shah Massoud in the North (Ghufran 2001, Hamilton et al.2001b, Harpviken 1997, Khalilzad 1995, Magnus 1997, 1998, Rubin 1997, Thaler et al.2010). Iranian assistance to Shi'a militias against the Sunni Taliban was a widely supported action within Iran as a way to protect co-religionists.

The political game in this era was especially complex. The main factions of the Mujahidin (namely the Northern Alliance) were supported by India, Russia, Iran, and Turkey. Only the motives of Turkey were not clear in getting involved in Afghanistan. On one hand Turkey and Iran frequently disputed one another. The sources of dispute included Turkey's close ties to the United States, its increased relationship with Israel in the 1990's, and its support for Azerbaijan in accordance to the role of the United States in Azerbaijan. The supporters and funders of the Taliban were Pakistan, Saudi Arabia, and the United Arab

Emirates (Ghufran 2001, Harpviken 1997, Khalilzad 1995, Magnus 1997, 1998, Rubin 1997). The only reasonable reason for support from Saudi Arabia and the United Arab Emirates was to curb Iran's influence in Afghanistan and to make sure that another Shi'a country would not emerge. The behind-the-scenes role of the United States resembles that of Great Britain's role in the region during the seventeenth through the nineteenth centuries. Although the United States did not support the Taliban overtly, its close allies were supporting opposing sides of the conflict in Afghanistan (Harpviken 1997, Ghufran 2001, Khalilzad 1995, Magnus 1997, 1998). Saudi Arabia, United Arab Emirates, and Pakistan were supporting the Taliban with funds and ammunitions, while Turkey was helping the Northern Alliance. A reasonable explanation for such behavior is that the United States preferred instability in Afghanistan until a friendlier government came to power. Furthermore the anti-Iranian Taliban were engaged in a struggle with Iranian backed Shi'a militias. Tacit support for the Taliban by the United States was intended to be an irritant to the Iranians.

In 1996, the Taliban ousted President Burhanuddin Rabbini after capturing Kabul and soon was recognized by Pakistan as the government in Afghanistan. In May 1997 the Taliban lost Mazar-i-Sharif to Northern Alliance forces and in June 1997 the Taliban closed the Iranian Embassy in Kabul in retaliation (Ghufran 2001, Harpviken 1997, Magnus 1997, 1998). In August 1998 United States Embassies in Kenya and Tanzania were bombed and Osama Bin Laden, the leader of Al Qaeda and a close partner of the Taliban, was accused of master-minding the attack. In the same month the Taliban declared they would support and protect Osama Bin Laden. The Taliban managed to single handedly dash the American dream of having an ally in Afghanistan. The Taliban began establishing a non-democratic government in Afghanistan and started the process of eliminating other factions and groups in the country (Ghufran 2001, Magnus 1997, 1998) Once again the United States found itself in a difficult position. On one hand it welcomed a strong central government in Afghanistan, especially since it was fighting and eliminating groups sympathetic to Iran or funded by Iran. On the other hand it could not stomach the non-democratic government in Afghanistan which was gradually forming by the Taliban's advance towards the north and capture of the majority of the country.

In August when the news of the massacre in Mazar-i-Sharif surfaced, Ayatollah Khomeini stated that the Taliban was seeking Iran's destruction. Tension between Iran and the Taliban escalated. When, in September 1998, the Taliban captured Mazar-i-Sharif and entered Iran's consulate they killed nine Iranian diplomats including the Consulate General. According to some estimates thousands of Hazaras, a minority group in Afghanistan who are also Shi'a, were massacred by the Taliban. Iran condemned the act and demanded an apology and retributions for murdering Iranian diplomats. The Taliban responded that it was not their fault and a rouge faction was responsible (Hamilton et al.2001b, Ghufran 2001, Thaler et al.2010). Iran, which already had some 70,000 troops at the border with Afghanistan, began amassing more troops at the border (Hamilton et al.2001b, Ghufran 2001, Thaler et al.2010). The United States became alarmed. If the presence of a pro-

Iran power in Afghanistan was worrisome for the United States one can imagine the level of anxiety if Iranian troops were actually present. The Islamic Republic of Iran, which had spent almost half of its existence fighting a difficult war, was reluctant to plunge into another war (given that country had managed to defeat the Soviet Union, not to mention Great Britain when in its prime). The tension escalated further when after a skirmish between the Taliban and Iran, which was allegedly started by a Taliban attack on an Iranian border post.

President Khatami proposed to take the matter to the United Nations. Kofi Anan, UN Secretary General, created a 6+2 group to address Iran's complaint and to avoid another regional conflict (Abrahamian 2008, Ghufran 2001, Hamilton et al.2001a, Hamilton et al.2001b, Thaler et al.2010). Iran's suggestion of a United Nations investigation was welcomed by United States. This situation brought another unique opportunity for normalization of relationships between the United States and Iran. Europeans were pushing for reduction, if not complete elimination, of sanctions against Iran. Iran has indicated through diplomatic channels in Europe and through the United Nations that they are willing to come to terms with the United States. An Iranian delegate was in New York for the United Nations general assembly. The Foreign Minister of Iran was to attend a meeting in which Secretary of State Madeline Albright was seated across the table from the Iranian delegate. She tried to break the ice and engage in a high level of diplomacy not realizing that the person that was seated at the head of the delegation was not the Foreign Minister of Iran but rather his deputy Mohammad Javad Zarif. Kofi Annan realizing the misunderstanding addressed the head of the Iranian delegation by his correct title. Needless to say the talks did not go anywhere. Apparently, this time it was Iran that played the United States and sent the wrong signal leading United States to believe that a new era was to be started. Nevertheless, the United States used its influence through allies to convince the Taliban to lower the level of hostility in the situation. This lost cause extended the conflict between Iran and the United States a few more years and reflects the sentiments of the American government and in turn the average American voter

The next opportunity for cooperation came in 2001 in a most unfortunate and sad form. On September 10, 2001 Ahmed Shah Massoud granted an interview with two reporters who had Belgium passports. A few minutes into the interview the camera man blasted explosives that he carried in his battery belt. The result of the explosion was the death of the Mujahidin leader. Alarm bells went off everywhere from Germany to Russia to India, to Iran and ultimately the United States. The consensus was that only Al Qaeda could pull something like this and that, based on their history, they most likely were about to attempt a bigger stunt. The saga of September 11, 2001 deserves a book of its own. For a brief moment the eyes of the world turned to Iran but immediately it was announced that Al Qaeda was responsible for the brutal attack. In fact the President of Iran and other dignitaries sent their condolences. All the anti-American demonstrations were halted and in their place public mourning of the victims popped up across Iran. By October 7 of the same year, the United States began attacking the caves and hide outs of Al

Qaeda in Afghanistan. The response came in the form of nightly bombing raids on major cities in Afghanistan where Al Qaeda operatives were believed to be present, as well as mountain ranges where the United States was aware of existing caves and hideouts (Quosh 2007).

Attacks on the Taliban were great news for Iran. President Khatami was not criticized for sending his condolences to the United States. Therefore, he took a bold step and suggested to help the United States in its war against the Taliban. This suggestion was not blocked either. Thus Iran began participating in low level intelligence cooperation with the United States against the Taliban working through the Office of the Secretary of State (Green et al.2009, Thaler et al.2010). At this time Khatami was locked in a battle with hard-liners and sought a diplomatic opening to strengthen his position and the position of the reformers. However, American sentiment was against Iran even though there was no evidence of Iranian support for Al- Qaeda.

The American's bombing campaign against the Taliban was not successful. Iranian military officers consistently suggested that the roadblock and positions of the Taliban surrounding Northern Alliance fighters must be bombed to free Northern Alliance fighters from their trapped positions. Finally, when with a nod from Colin Powel the Secretary of State, the Americans targeted Taliban positions besieging the Northern Alliance fighters (Quosh 2007). Consequently, these foot soldiers managed to join the fight against the Taliban. With their knowledge of the terrain and their contacts in the population, the position of the Taliban weakened. After this gesture, Iran expected the United States to reciprocate. In fact Secretary of State Powell was in favor of showing some gesture of goodwill on the part of the United States. The Secretary of State's opinion was not shared by President George W. Bush's White House advisors. For example, John Bolton had stated for the record that "Iran must get down on their knees and thank us for getting rid of their enemy" (Mokhtari 2005, Quosh 2007). Hence another opportunity for normalizing the relationship between the two countries was lost. This really hurt the position of President Khatami, who had campaigned on the platform of "dialogue of civilizations." Both before and after election to office Khatami stated his willingness to have a dialogue with the United States as equals with mutual respect. Khatami's extension of friendship came at an inopportune time. The American government was in no mood to open a dialogue. The failure to respond to Khatami's overtures is a reflection of the sentiment of the American selectorate. Moreover, entering talks with Iran would have been seen by the American public as a sign of weakness on the part of the administration, something the president wanted to avoid. Using Iran as the "Bad Guy" when possible is a reflection of the continued conflict between the two nations. When either government needs a "Bad Guy" to blame for whatever issue is at hand, they have a readily available "Bad Guy" in the other.

Furthermore, the September 11 attack had allowed a much harder stance on America's foes. Neither allies of the United States nor opponents wanted to be on the wrong side of the conflict. In fact President George W. Bush in a November

6, 2001 speech declared to the world that "either you are with us or against us in the fight against terror." The United States government took advantage of the attack to demand and obtain much more power to search, siege, and arrest at home and to attack Taliban and Al Qaeda in Afghanistan. President George W. Bush went further and unilaterally stated that the United States will attack any terrorist group and/or their supporters anywhere in the world any time without further provocation or direct attack on the United States or its interests (Mokhtari 2005, Quosh 2007). Understandably, it was in no one's interest to challenge the United States in light of the argument of self defense. Furthermore, the United States government claimed that Al Qaeda is seeking the so-called "dirty bomb," which is a conventional bomb laced with the radioactive (possibility waste) materials. The possibility of a plane load of explosives with massive amounts of radioactive materials exploding above a large city in United States, or Europe for that matter, made every government and official nervous, not to mention ordinary people residing in such large cities. Additional fear was spread due to the fact that, although some larger cities could have been protected by their respective nation's air forces, there was no realistic way of protecting medium size cities with populations of several hundred thousand.

Soon, there were American troops on the ground in numerous parts of Afghanistan. To solve logistical problems and to strengthen its military maneuvers, the United States strengthened existing agreements and secured new ones with Azerbaijan, Uzbekistan, Kyrgyzstan, Turkmenistan, Pakistan, Qatar, Bahrain, Kuwait, Saudi Arabia, and Israel. Iran, which was uncomfortable with the US presence in Persian Gulf states, found itself surrounded by US military on many sides, within striking distance of its bombers. To make the matters worse, almost everyone was certain that the United States was preparing to attack Iraq. The main reason for such anticipation was the increased accusation by the United States that Iraq was producing weapons of mass destruction that it could and would share with Al Qaeda. Iraq's cat and mouse game with inspectors and tough talk was another reason for such predications. Once again, Iran offered to assist the United States, but this time in Iraq. Iran suggested a 6+ 6 formula in place of the previously successful 6+2. The latter group would consist of the five neighboring countries of Iraq plus Egypt (Cordesman & Al-Rodhan 2006, Green et al.2009, Mokhtari 2005, Thaler et al.2010).

It seemed that the United States was not interested in a diplomatic solution for Iraq's lack of cooperation on the weapons of mass destruction. It could be that the United States saw this as a sign of Iran's weakness and/or worry. Furthermore, it seemed that the United States was determined to attack Iraq and did not mind the possibility that Iran could have been concerned for its own security. The sentiment was not limited to Iranians. If there were any doubts about how the United States felt towards Iran President George W. Bush's January 29 speech in which it named North Korea, Iraq, and Iran the Axis of Evil clarified those doubts (Mokhtari 2005, Quosh 2007, Thaler et al.2010). In fact the talks of attack on Iran skyrocketed. The belief was that as soon as the United States had enough troops in Kuwait and Saudi

Arabia it would start a massive aerial attack on Iraqi positions. The Shock of Awe attack would soften Iraqi troops and then the rest of the American forces would move in for the final kill. The sentiment was that Iraq, although much larger and stronger than Afghanistan, would fall soon similar to the situation in Afghanistan. The scenario continued, claiming that at that time the United States would be in position to attack Iran from numerous fronts and change the government. The United States might have been too anxious to show that it was extremely strong and could win regional wars rapidly and easily and on May, 2003 President George W. Bush arranged for a photo opportunity on the deck of the USS Abraham Lincoln, arriving on a jet and declaring "mission accomplished."

All the talks about the Axis of Evil, and the regime change in Iran, had lead to speculations that as soon as the United States won the war in Iraq and Afghanistan, it would begin positioning its troops to attack Iran. At this time the United States had troops to the east, south, and west of Iran and very close ties with countries to the north of Iran (in central Asia) with military cooperation agreements. With the declaration of "mission accomplished" the speculations intensified. All this made many European countries worried that an attack on Iran was imminent. Due to internal pressure from their citizens, the European countries were stretched to their limits and did not want to be part of another United States offensive in Central Asia. Therefore, under leadership of Germany and France (and later England) a chorus was forming against military attack on Iran.

The reality was that the war was far from over in either Afghanistan or Iraq. For all practical purposes, the United States had lowered its efforts in Afghanistan expecting its allies to maintain order and support President Karzi's regime in establishing a stable government. The Taliban was gradually reestablishing its power bases in the country. In fact many years later when President Obama was campaigning, he promised to increase United States presence in Afghanistan and remove the Taliban from the country. Situations in Iraq also began deteriorating. Shi'a factions had chosen to both flex their muscles and take parts of the country under their control and at the same time engage in a grassroots campaign to become a political power in a democratically elected government at every level. On the other hand, Sunni groups had succeeded in creating a reign of terror through numerous bombs targeting crowded areas in major cities and where killing Shi'a civilians everywhere from shopping centers to Mosques.

Within the same time period the possibility of nuclear proliferation through Iran took center stage and issues of assisting Hamas or Hezbollah became secondary. Talks of regime change became less frequent and the United States toned down its opposition to Iran. The US even offered a role for Hezbollah in Palestinian issues (and in Lebanon) if they stopped their opposition to Israel's right to exist and stopped guerrilla attacks on Israel and in Lebanon. Germany, France, and England (EU3) began shuttle diplomacy with Iran in order to preempt a US attack on Iran (Cordesman & Al-Rodhan 2006, Kibroglu 2006, Quosh 2007, Thaler et al.2010). At times it seemed that Iran would agree with all EU3 demands regarding a stop of all nuclear research. At other points during negotiations it would seem as though

Iran was in total defiance. The main obstacle was Iran's insistence that it should be allowed to research uranium enrichment. In return Iran would halt building centrifuges (Cordesman & Al-Rodhan 2006, Crane et al.2008, Kibroglu 2006, Quosh 2007, Thaler et al.2010).

Although Iran's nuclear program can be traced to the 1950s, it was only in the mid 1970s that Iran actually began a serious program (Abrahamian 2008, Cordesman & Al-Rodhan 2006, Green et al.2009, Hamilton et al.2001b, Kibroglu 2006, Thaler et al.2010). Then, however, the West was eager to help and did not see any threat. France managed to get the lion's share of the contracts but the United States was eager to participate as well. The Iranian Revolution put an end to all that. The nuclear program had further setbacks when Iraq invaded Iran and the war depleted all the resources from everywhere, including nuclear programs. Further impediments came in the form of Iraqi bombs that damaged the main nuclear facility in Bushehr (Abrahamian 2008, Cordesman & Al-Rodhan 2006, Green et al.2009, Hamilton et al.2001b, Kibroglu 2006, Thaler et al.2010). Once the war was over and Iran could get back on its feet, it began reestablishing its nuclear program. In pursuit of that, Iran contacted France, Argentina, Russia, North Korea, China, and Pakistan to name few. In return, the United States pressured each country one-by-one to ensure they did not assist Iran. Eventually it was Russia, motivated by its own interests, that began rebuilding the damaged Bushehr plant. It turned out that the cooperation was more than just rebuilding a damaged nuclear plant. It involved training, the transfer of technology, the provision of advisors, and information on how to build guided missiles (Abrahamian 2008, Cordesman & Al-Rodhan 2006, Green *et al.*2009, Hamilton et al.2001b, Kibroglu 2006, Thaler et al.2010, Wehrey et al.2009). In July 2010 Russia began fueling the Bushehr reactor igniting speculation that Israel would attack the facility. No attack materialized but speculation that the facility could be used to produce plutonium for nuclear weapons continues. The starting of the reactor was a victory for the hardliners, who sought to defy world opinion and especially the wishes of the United States.

When the United States could not sway Russia it began pressuring Iran and demanded an end to its nuclear weapons program. In response, in 1992, Iran offered to allow the International Atomic Energy Agency (IAEA) to have unrestricted access to all nuclear facilities in Iran (Cordesman & Al-Rodhan 2006, Hamilton et al.2001b, Kibroglu 2006, Mokhtari 2005, Quosh 2007, Thaler et al.2010). Consequently, the agency declared that, after examining all Iranian facilities, Iran's claim of the peaceful use of nuclear power was substantiated and there was no evidence of any nuclear bomb program in the country (Cordesman & Al-Rodhan 2006, Hamilton et al.2001b, Kibroglu 2006, Mokhtari 2005, Quosh 2007, Thaler et al.2010). European countries celebrated the news and agreed to reduce their sanctions and increase ties with Iran. However, the United States refused to validate the report or change its stance with Iran. Finally, Argentina (under continued pressure from United States) canceled its nuclear agreement with Iran.

In August 2002 the media reported allegations of a uranium enrichment plant located in the central city of Natanz and a Heavy Water plant in vicinity of the capital in the industrial city of Arak (Cordesman & Al-Rodhan 2006, Crane et al.2008, Kibroglu 2006, Mokhtari 2005, Quosh 2007, Thaler et al.2010, Wehrey et al.2009). However, it did not bring a big response from the government of United States, which apparently knew about both facilities and had even discussed the possibility of using bunker penetrating bombs against the former. Once again Iran offered full access to the IAEA as a gesture of goodwill but demanded a gesture of goodwill in return. Specifically President Khatami requested civilian airplane parts for the aging American built fleet which were crashing with alarming frequency (Abrahamian 2008, Cordesman & Al-Rodhan 2006, Hamilton et al.2001a, 2001b, Kibroglu 2006, Mokhtari 2005, Quosh 2007, Thaler et al.2010). In 2004 the IAEA inspected the new plants and found no evidence of military related nuclear activity (Cordesman & Al-Rodhan 2006, Kibroglu 2006, Quosh 2007, Thaler et al.2010, Wehrey et al.2009).

EU3 negotiators were pleased and created a proposal whereby the Iranians would agree to continued inspection and the Americans would provide spare parts for the planes. The Office of Secretary of State had given a green light and stated that the proposal was agreeable for the Americans. When the EU3 delegate arrived with the document instead of the Colin Powell, the Secretary of State, they were received by John Bolton, who had been the Undersecretary of State for Arms Control and International Security since 2001 (Quosh 2007). This was not good news, especially in light of Bolton's reputation. Bolton, a conservative, frequently protested against regimes that were considered unfriendly. In 2002 he accused Cuba of creating weapons of mass destruction (Quosh 2007). He also had accused both Iran and Syria of making and owning weapons of mass destruction and had declared them threats to regional stability. The subsequent Israeli attack on a "research" facility in Syria seems to indicate that the Assad regime was seeking some sort of WMD capability.

Allegations about his fabrication of facts to make and support his points were abundant. Later, his nomination for the post of United States Ambassador to the United Nations was delayed because of a Congressional hint that he would not be confirmed. Consequently, President George W. Bush appointed him to the post while Congress was in recess. When it was time for an official vote for continuation in the post Bolton resigned. Bolton stated that the United States would not agree to any of terms of the negotiations, declared Iran a sponsor of terrorism, and iterated the desire for a government change in Iran (Quosh 2007, Torbat 2005). Consequently, there was no deal.

This was the final blow to the moderate government of President Khatami and hardliners in Iran began consolidating their forces and the President was marginalized (Abrahamian 2008, Green et al.2009, Thaler et al.2010, Torbat 2005). The fruits of this series of events soon appeared in the form of nominees for presidential election in Iran. What resulted was victory for hardliner candidates, namely President Ahmadinejad and the hard-liners. The election of Ahmadinejad indicates how the elites in the Iranian selectorate have the ability to determine who runs and who wins elections. Moreover, it also indicates that the selectorate in Iran is shrinking from the general population to elites who control the armed-forces, Pasdaran, the bureaucracy and the Clergy. This consolidation of power makes it more difficult for reformers to gain influence.

In February 2005, Condolezza Rice became the Secretary of State. She was a close friend of President George W. Bush and philosophically followed the same values. She realized how few resources were placed on studying Iran and analyzing the situation in Iran. This was a result of the worsening of the situation in Iraq. American casualties were mounting due to powerful roadside bombs referred to as Improvised Explosive Devices (IED). The United States blamed Iran for providing the new improved explosive devices that could actually penetrate the heavy armor of many American vehicles (Crane et al.2008, Quosh 2007, Wehrey et al.2009). At the same time the bombing of civilians in crowded city dwellings generated fear among the civilians who could not even shop for life necessities such as food. Both England and United States blamed Iranians for providing the IEDs to Iraqi militants. In response, Iranians denied all accusations pointing to the fact that tons of explosives had been stolen in Iraq, which were utilized in making the home-made bombs. Another source for these bombs is the unexploded American ordinance. The use of unexploded ordinances dates back to the Vietnam War. Iranians improved the technology during the war between Hezbollah and Israel in Southern Lebanon in 2000. American and British governments continued warning and threatening Iran with more sanctions to outright invasion. When there were no doubts in the minds of both sides about who was providing the technology, and even the equipment and how effective they were, the Iranians approached EU3 diplomats and offered to stop killing Coalition forces in return for lifting of the embargo against Iran (Crane et al.2008, Quosh 2007, Wehrey et al.2009). Minor agreements were reached and Iran was given a greater role in Iraqi negotiations but nothing major came out of this. The United States implied that "moderate" elements in the Iranian government will be supported and helped to come to power. This was a much softer stance than the outright regime change of years past. The United States, however, was still involved in numerous provocations and activities against Iran. The United States supported dissident Iranian groups which were based in Iraq such as the Mojahedin Khalgh (MKO) as well as the Party of Free Life in Kurdistan (PEJAK). These groups infiltrated Iran; enticed violence; and tried to flare up ethnic and ideological upraise. On several occasions American forces in Iraq crossed the border and entered Iran. The most contested activity, however, was and still is the flight of Drones, other surveillance aircraft, and satellites over Iran.

The election of President Ahmadinejad demonstrated that hardliners had succeeded in consolidating their power (Abrahamian 2008, Crane et al.2008, Green et al.2009, Hamilton et al.2001a, Thaler et al.2010). Consequently, the hopes for any reconciliation with the West, let alone the United States, were dashed. By now the United States had massed numerous troops in Iraq and Afghanistan. The

Americans established logistic infrastructure in both countries as needed and had completed many scenarios for invading Iran.

When President George W. Bush was swept into the office for the second time, he interpreted re-election as the support of Americans for his domestic and foreign policy. The administration interpreted the result as a mandate to fight terrorism. Soon after election he stated that he had gained substantial political capital and he intended to spend it. One interpretation was that the President had seen this as the green light to attack Iran and possibly North Korea, if not both. It seemed that the election of a hard-line president in Iran who immediately restarted the Iranian uranium enrichment program was the excuse that President Bush needed. The Republican president felt that the selectorate who was behind the initial invasion of Afghanistan and Iraq may be willing to confront Iran. However, as causalities mounted in Iraq the selectorate turned against the war-effort and President Bush.

Just as with Iraq, the United States orchestrated a set of activities. It contacted the United Nations and the IAEA accusing Iran of seeking a nuclear bomb. The American media was riddled with accusations of Iran's imminent access to nuclear bombs, with differing dates for actual detonation of a device. Simultaneously the talks of state-sponsored terrorism flared up. At times Iran was grouped with North Korea, at other times it was bundled with Syria, and occasionally Libya was added to the mix. Needless to say Israel, Hamas, Hezbollah, and terrorism were also a common reference whenever Iran was addressed. Of course President Ahmadinejad's references to the Holocaust or declaration that Israel will be "wiped off of the map" not only did not help but also provided fuel to the frenzy of talks about attacking Iran.

Although, at the surface, it seemed that the United States was going to attack Iran, the attack did not take place. This was due in part to European and other allies informing the United States that if Iran was going to be attacked that the United States would do so alone. Obviously, other leaders did not feel they have any "political capital" to spare. Public opinion polls in country after country revealed that the United States was seen as a bully, not a savior of the world. Much of the world including a sizable portion of Americans, were not in favor of an American presence in Iraq and Afghanistan. The support for a United States invasion of Iraq was 75% in 2003 according to a USA Today / Gallop poll ("Wartime Dissent" 2005). Support eroded, however, as the war was prolonged and truth became known about the alleged Iraqi weapons of mass destruction. According to numerous polls in 2007, about 61% of Americans believed that the United States "should have stayed out of Iraq." Needless to say, the opinions in other countries were less favorable. In short, the United States did not attack Iran in spite of the more defiant stance Iran assumed after President Ahmadinejad's election. Iran's resumption of uranium enrichment and testing of mid-range ballistic missiles did not trigger a war. This episode provides another example of how domestic politics in both countries can drive policy and how both nations' policy makers use the image of the "Bad Guy" to garner support for their policies.

When it became clear that the United States would be alone if it was to attack Iran (and even its closest ally, England, would not join) and that the possibility of a United Nations' resolution would not transpire, the war talk subsided. In fact, Secretary of State Rice publicly announced that the United States would negotiate with Iran as equals, with respect, and with everything on the table (Quosh 2007, Thaler et al.2010, Wehrey et al.2009). She continued with the tough talk as well, threatening that if Iran refused to participate in negotiations, continued sponsoring terrorist groups, and continued with its nuclear ambitions then there would be consequences. However, few considered this the same as the warnings preceding the invasion of Iraq. It was apparent that the United States had softened its tone against Iran due to the reality of international relations. Though talks and treats did not return until after President Obama's second year in office. Iranians anticipated a dialogue with the United States and normalization of diplomatic relations. Not only were all the major European powers encouraging Iran to take advantage of the opportunity, but all neighboring countries and regional powers were telling Iran that the offer was a great one and Iran should reciprocate the effort.

The biggest change was that there was no more talk of regime change in Iran. The main focus was to deter Iran from pursuit of a nuclear bomb. To that end an offer came from United States via Javier Perez de Cuellar. The offer, which was backed by European powers, allowed Iran to continue centrifuge research as long as uranium enrichment was halted. This was a major shift in the United States's position that had originally opposed any nuclear activities, research, or actual enrichment. The offer took the Iranians by surprise. Iran has been insisting that its nuclear program is for peaceful purposes and it needs the technology to meet the country's energy needs when their oil runs out. Iran believes it should be allowed to conduct research and improve its centrifuges. The negotiations asked for simultaneous temporary actions by both Iran and the United States. Iran would suspend enrichment, to be verified by free inspections from the IAEA, and the United States would lift sanctions. If both parties were satisfied that the terms of agreement were met, an official and permanent agreement would be signed. It seemed that the offer was what the Iranians wanted all along. This aroused suspicion in Iran. Some leaders, especially President Ahmadinejad, considered this a trap and pulled back. As a consequence another major opportunity was missed to normalize relations between the two countries.

There are several conclusions that can be drawn from this examination of Iranian American relations. Politicians in both countries tend to use the other as the preverbal "Bad guy" and paint them as an adversary, even if this is not the case. This sort of action is a tactic to rally support from the selecotrate to help them maintain their political office or in the case of Iran for the hard-liners to consolidate their power after the moderate Khatami's presidency. The reliance on conflict imagery when dealing with the other country allows decision makers to leverage the past history of low-level conflict to their advantage. Thus, the conflict continues on the economic, political and military spheres of action.

Bibliography

Abrahamian, E., 2008. *A History of Modern Iran*, NYC: Cambridge University Press.Byman, D., 2003. Should Hezbollah Be Next? *Foreign Affairs*, 82(6), 54-66.

Behrooz, M., 2001. Tudeh Factionalism and the 1953 Coup in Iran. *International Journal of Middle Eastern Studies*, 33, 363-382.

Chaqueri, C., 1999. Did the Soviets Play a Role in Founding the Tudeh Party in Iran? *Cahiers du monde Russe*, 40(3), 497-528.

Cooper, A. & Telfer, L., 2006. Misperceptions and Impediments in the US-Iran Relationship. *49th Parallel*, 18(Summer), 1-42.

Cordesman, A.H. & Al-Rodhan, K.R., 2006. The Gulf Military Forces in an Era of Asymmetric War: Iran. *International Studies*, 1(202).

Cordesman, A.H., 2006. *Iran's Support of the Hezbollah in Lebanon*, Washington, D.C.

Cordesman, A.H., 2007. *Iran's Revolutionary Guards, the Al Quds Force, and Other Intelligence and Paramilitary Forces*, Washington, D.C.

Crane, K., Lal, R. & Martini, J., 2008. *Iran's Political, Demographic, and Economic Vulnerabilities*, Santa Monica, CA: RAND.

Daniel, E.R., 2001. *The History of Iran*, Westport, CT.: Greenwood Press.

Danielsen, A.L., 2010. OPEC. *Encyclopaedia Britannica Online*. Available at: http://search.eb.com.logon.lynx.lib.usm.edu/eb/article-233528.

Dewdney, J.C., 2010. Union of Soviet Socialist Republics. *Encyclopaedia Britannica Online*. Available at: http://search.eb.com.logon.lynx.lib.usm.edu/eb/article-9105999.

Duby, G., 1987. *Atlas Historique Duby*, Paris: Larousse.

Duiker, W.J. & Turley, W.S., 2010. Vietnam. *Encyclopaedia Britannica Online*. Available at: http://search.eb.com.logon.lynx.lib.usm.edu/eb/article-52748.

Erb, H. J., & Reuter, L. R. (2010). Berlin. *Encyclopaedia Britannica Online*. Encyclopaedia Britannica. Retrieved from http://search.eb.com.logon.lynx.lib.usm.edu/eb/article-21658.

Ghufran, N., 2001. The Taliban and the Civil War Entanglement in Afghanistan. *Asian Survey*, 41(3), 462-487. Available at: http://caliber.ucpress.net/doi/abs/10.1525%2Fas.2001.41.3.462.

Green, J.D., Wehrey, F. & Wolf, C.J., 2009. *Understanding Iran*, Santa Monica, CA: Rand.

Hajjar, S.G., 2002. *Hizballah: Terrorism, National Liberation, or Menace?*, Carlisle, PA: Strategic Studies Institute.

Harpviken, K.B., 1997. Transcending Traditionalism: The Emergence of Non-State Military Formations in Afghanistan. *Journal of Peace Research*, 34(3), 271-287.

Hicks, B.D., 1996. Presidential Foreign Policy Prerogative after the Iran-Contra Affair: A Review Essay. *Presidential Studies Quarterly*, 26(4), 962-977.

Jordet, N., Explaining the Long-term Hostility between the United States and Iran : A Historical, Theoretical and Methodological Framework. *Thesis*

Khalilzad, Z., 1995. Afghanistan in 1994: Civil War and Disintegration. *Asian Survey*, 35(2), 147-152. Available at: http://caliber.ucpress.net/doi/abs/10.1525/as.1995.35.2.00p0469f.

Khalilzad, Z. & Byman, D., 2000. Afghanistan: The Consolidation of a Rogue State. *Washington Quarterly*, 23(1), 65-78. Available at: http://www.mitpressjournals.org/doi/abs/10.1162/016366000560746.

Kibroglu, M., 2006. Good for the Shah, Banned for the Mullahs: The West and Iran's Quest for Nuclear Power. *Middle East Journal*, 60(2), 207-232.

Kinsella, D., 1994. Conflict in Context: Arms Transfers and Third World Rivalries during the Cold War. American Journal of Political Science, 38(3), 557-581. Available at: http://www.jstor.org/stable/2111597?origin=crossref.

Magnus, R.H., 1997. Afghanistan in 1996: Year of the Taliban. *Asian Survey*, 37(2), 111-117. Available at: http://caliber.ucpress.net/doi/abs/10.1525/as.1997.37.2.01p0209h.

Magnus, R.H., 1998. Afghanistan in 1997: The War Moves North. *Asian Survey*, 38(2), 109-115. Available at: http://caliber.ucpress.net/doi/abs/10.1525/as.1998.38.2.01p0322p.

Meskill, C.M., 1995. American Diplomacy in the Iranian Revolution, 1976-1981.

Mistry, D., 2003. Beyond the MTCR: Building a Comprehensive Regime to Contain Ballistic Missile Proliferation. *International Security*, 27(4), 119-149. Available at: http://www.jstor.org/stable/2606509?origin=crossref.

Mokhtari, F., 2005. No One Will Scratch My Back: Iranian Security Perceptions in Historical Context. *Middle East Journal*, 59(2), 209-229.

2010. Persian Gulf War. *Encyclopaedia Britannica Online*. Available at: http://search.eb.com.logon.lynx.lib.usm.edu/eb/article-9059340.

Quosh, C., 2007. *American Foreign Policy Towards Iran: Between Values and Interests or Beyond?*, Hamburg.

Rakel, E.P., 2009. The Political Elite in the Islamic Republic of Iran: From Khomeini to Ahmadinejad. *Comparative Studies of South Asia, Africa and the Middle East*, 29(1), 105-125. Available at: http://cssaame.dukejournals.org/cgi/doi/10.1215/1089201X-2008-047.

Renton, A.W., 1933. The Revolt Against the Capitulatory System. *Journal of Comparative Legislation and International Law, Third Series*, 15(4), 212-231.

Roberts, M., 1996. *Khomeini's Incorporation of the Iranian Military*, Washington, D.C.

Rubin, B.R., 1997. Women and Pipelines : Afghanistan's Proxy Wars. *International Affairs*, 73(2), 283-296.

1912. Russia and Persia. *The American Journal of International Law*, 6(1), 155-159. Available at: http://www.jstor.org/stable/2187403?origin=crossref.

Schahgaldian, N.B., 1989. *The Clerical Establishment in Iran*, Santa Monica, CA: Rand.

Smith, L., 2007. *Iran, Hizbullah, Hamas, and the Global Jihad: A New conflict Paradigm for the West* L. Smith, Jerusalem: Jerusalem Center for Public Affairs.

Spector, R. H. (2010). Vietnam War. *Encyclopaedia Britannica Online*. Encyclopaedia Britannica. Retrieved from http://search.eb.com.logon.lynx. lib.usm.edu/eb/article-234639.

Tapp, J., 1951. The Soviet-Persian Treaty of 1921. *The International Law Quarterly*, 4(4), 511-514. Available at: http://www.journals.cambridge.org/ abstract_S0020589300007545.

Thaler, D.E., 2010. *Mullahs, Guards, and Bonyads: An Exploration of Iranian Leadership Dynamics*, Santa Monica, CA: Rand

Torbat, A.E., 2005. Impacts of the US Trade and Financial Sanctions on Iran., 407-434.

"Wartime Dissent." *Issues & Controversies On File:* n. pag. *Issues & Controversies*. Facts On File News Services, 7 Oct. 2005. Web. 29 June 2010. <http://www.2facts.com.logon.lynx.lib.usm.edu/article/i1000530>.

Wright, R. & Bakhash, S., 1997. The U.S. and Iran: An Offer They Can't Refuse? *Foreign Policy*, 108, 124-137.

Wehrey, F., 2009. *Dangerous But Not Omnipotent: Exploring the Reach and Limitations of Iranian Power in the Middle East*, Santa Monica, CA: RAND.

Wehrey, F., 2009. *Saudi-Iranian Relations Since the Fall of Saddam: Rivalry, Cooperation, and Implications for U.S. Policy*, Santa Monica, CA: RAND.

Wehrey, F., 2009. *The Rise of The Pasdaran: Assessing the Domestic Roles of Iran's Revolutionary Guards Corps*, Santa Monica, CA: RAND.

Wonnacott, P., 2010. Commodity Trade. *Encyclopaedia Britannica Online*. Available at: http://search.eb.com.logon.lynx.lib.usm.edu/eb/article-8673.

Chapter 7

Low-Level Military Confrontation in the Persian Gulf

This chapter examines the low-level military confrontation that has occurred between Iran and the United States in the past thirty years. Over the past thirty years, the ebb and flow of the conflict has produced actual shooting and death. At other times the conflict has produced long periods of non-violent yet intense hostility. This chapter will examine several of these periods and explore the veracity of the conflict, while looking at the underlying political forces that drove the various phases of the overall Iranian-American low-intensity war. The first is the period encompassing the hostage crises and the Iran-Contra affair. The second is the Iran-Iraq War and the Tanker War, and the third is the ongoing conflict over the Iranian Nuclear Weapons program. This chapter examines these periods with an eye toward how the internal politics of each nation drives policy decisions and continues the low-level conflict between the two countries.

Hostage Crises and Iran-Contra Affair

The fall of the Shah of Iran marked a dramatic turning point in Iran-US relations. The new regime that coalesced around Ayatollah Khomeini was fundamentally hostile to the United States and American interests in Iran. The revolutionary fervor of the times, along with the need for Khomeini to solidify his grasp on power and institutionalize his revolutionary government, made the taking of the American embassy feasible for the radical student groups who did indeed take the embassy on November 4, 1979. While the Ayatollah never publicly called for or acknowledged prior knowledge of the takeover, he tacitly approved of the action. As a result of the hostage crisis the government of Iran refocused attention on American wrongs and the Shah who was supported by the United States for years and had been allowed medical treatment in American hospitals. Interestingly, this was not the first time a foreign embassy had been taken. In 1829 the Russian Embassy was attacked by a mob, the diplomatic guards were killed, and the ambassador was killed then beheaded (Kaplan 1996: 186). While the 1979 event lasted longer, none of the hostages were killed and several were released early or for humanitarian reasons. Nevertheless, this incident put the United States and Iran on a collision course and essentially marked the beginning of the low-intensity conflict.

The embassy takeover also demonstrates how each country's selectorate works and how the coalition that backs the government can exercise its power. For Iran, the taking of the embassy resulted in breaking the deadlock which prevented the new Constitution's writing. The debate had revolved, among other things, around the rule of the jurist and the notion of a Supreme Leader with veto powers over all facets of government and social life. The taking of the hostages symbolized the power of the radicals who supported, and were supported by, Khomeini and his clerical revolutionaries. The symbolism reminded the population that not only had Revolutionary Islamic action taken down the Shah, but now it had the power to take American hostages and property with impunity. The humiliation of a superpower was not lost on the majority of the population who, while happy to see the Shah go, were hesitant to embrace the revolutionary agenda of Khomeini and the radical students. This activity demonstrated the strength of the Islamic radicals who were able to force through their preferences in the Assembly of Experts for the Constitution. The specific inclusion of the Supreme Leader as a constitutionally mandated position was added to the constitution and a referendum was held. Not surprisingly the referendum passed. However the turnout was small and the voting supervision by the radical militia was not a secret. Once the institutional arrangement and various powers of the institutions and actors were enabled, the hostages had less importance other than to obtain concessions from the United States. With their utility waning as revolutionary institutions consolidated power, two events shaped the release of the hostages. The first was the beginning of the Iran-Iraq war, or the "Imposed War" as it is called in Iran, and the election of Ronald Reagan to the White House.

Saddam Hussein's invasion of Iran allowed the new revolutionary government and political selectorate to consolidate their positions and to institutionalize the revolution. With all social and political groups intent on supporting the government and prosecuting the war any opposition was effectively muted. Despite the heavy price in human lives most segments of society supported the war efforts. Prior to the invasion the hostages (or Hostage incident) had served the function of mobilizing the revolutionaries for the new Islamic Constitution. Now they had a different function. Their new function, however, was short lived. The war put the hostages on the back burner for the regime as it attempted to defend its territory. The low rate of serviceability for the Iranian military following years of neglect allowed Iraq to make considerable gains initially. The regime was concerned that Iraqi technical superiority would turn the tide of battle. In Operation Morvarid on November 28, 1980 Iranian air force and navy units successfully defeated Iraqi naval and air units, destroyed oil terminals, oil installations, destroyed surface to air missile sites and also destroyed over 80% of Iraqi naval vessels (Cordesman & Wagner 1990, Mokhtari 2005, Pollack 2005, Sick 1989, Sterner 1984). While a victory, the loss of an F-4 Phantom fighter and the hasty repositioning of more advanced F-14 fighters demonstrated that spare parts of American built equipment were in short supply and thus limited offensive operations. While presumably the government would have liked to use the hostages in return for spares, this was not the case as Ronald Reagan was elected to the American presidency. Seeing the shift in the American stance toward Iran the hostages were released prior to

Reagan's inauguration. Arms for hostages would have to wait until 1985 and the Arms for Hostages deal.

The election of Reagan signaled a shift in the American electorate. Gone was the cynicism of the post-Nixon years, but the moral basis President Carter attempted to bring into American foreign policy was seen as a failure. The fall of the Shah, the Soviet Invasion of Afghanistan, the Panama Canal Treaty and the economy were major factors that lead to the Reagan landslide. Stunned by a perceived lack of US willpower or ability to do anything correct, the US electorate turned to Reagan's message of increased American strength, a tougher American foreign policy (especially toward the Soviets and Iran), and his willingness to use military force to back up diplomacy (Kelley 2007, Pollack 2005, Posen & Van Evera 1983, Stein 1989). The American selectorate choose a President that sought foreign policy victories to help the American populace regain its composure after the humiliation of the Hostage Crises and the unsuccessful rescues attempt.

The effect of Ronald Reagan's election was immediate on Iranian-American relations. Many Americans felt that Reagan's tough talk had intimidated the Iranians into releasing the hostages. Once the hostages were released, the administration treated the incident as an act of war, an attitude still harbored in many quarters of the American electorate to this day. The Iran-Iraq war took center stage with the United States oscillating between its support for both Iraq and Iran. A stalemate suited American interests as the war kept Iran engaged in a defensive war while containing Iraq's expansionistic tendencies. Given the preoccupation with the Soviets in Afghanistan, it was in American interests that Iran was preoccupied so they did not cause any mischief in the Gulf monarchies.

By 1983 the United States had several hostages taken by Islamic radicals in Lebanon, namely Hezbollah. With the death of the Lebanese Shiite leader Musa Al-Sadr, Lebanese Shi'a leaders moved from the moderate Amal political movement and militia to the more radical and confrontational Hezbollah. Following the bombing of the United States Marine Corps barracks in Beirut, American forces pulled out of Lebanon without having pacified the warring parties or liberating the hostages. In an effort to secure the release of the hostages, the Reagan administration began a program that would provide Iran with weaponry, and in return Iran would bring pressure on Hezbollah to release the American hostages. Clearly the Americans wanted all the hostages released so it could pursue a more aggressive Middle East policy free of potential political fallout from pictures of dead American hostages on television (Bogen & Lynch 1989, Brody et al. 1989, Canham-Clyne 1992, Cavender 1993, Hicks 1996, Koh 1988, Kornbluh 1988, Rubenberg 1988, Sharpe 1987). Iran, rationally knowing this, ensured that their Hezbollah allies did not release all the hostages, in order for the arms to keep arriving and to maintain some sort of trump card over the United States. Between 1984 and 1986 the United States supported Iran in its war and Iraq through shipments of various types of arms. If Iran were to repel the Iraqi offensive it would have to use its technological superiority in aircraft and surface to air weapons in the initial stages, which is what it did. The success of Operation Morvarid demonstrated that training Iranian pilots had undergone in the United States was superior to that of the Soviet trained Iraqi pilots. Compounding the Iraqi problem was the technological superiority of the American aircraft purchased by the Shah and used by the Islamic Republic. While having aerial superiority, Iran faced the serious problem of spare parts. The F-4 Phantoms and F-14 Tomcats (the most modern and effective Iranian aircraft) required significant preventative maintenance before each flight as well as subsequent post-flight maintenance. Critical parts for engines, electronic systems, and instruments have to be replaced regularly and in some cases very frequently.

Fully aware that Iran needed spares for its aircraft and replacements for some of its other weapon systems, the United States began a program that would later become known as the Iran-Contra affair. Due to congressional prohibitions on the funding of Nicaraguan insurgents known as the Contras, who were attempting to overthrow the Sandinista regime of Daniel Ortega, the United States needed an alternative source of funding. Covertly selling weapons to Iran and funneling the proceeds to the Contras was the illegal part of the scandal that prompted congressional hearings (Bogen & Lynch 1989, Cavender et al. 1993, Scheffer 1987).

The weapons that were sold (or given) to Iran were inherently of the defensive sort. Besides the need for spare parts for its aircraft which were delivered in small batches, arms were needed by the Iranian Army to fight the ground war against Iraq. Two major types of weapons were supplied to Iran between 1984 and 1986. These included TOW wire-guided anti-tank missiles which the Iranians had prior to the revolution. The other missiles were HAWK surface to air missiles that were intended to deter the numerically superior Iraqi air force. While the TOW missiles were used to blunt the Iraqi advantage in armored forces, the HAWK missiles were intended to defend the Iranian battle lines and cities from Iraqi aerial bombardment. Both weapon systems are inherently defensive in nature. Why would the United States limit the Iranian weapons deliveries to defensive weapons? The simple answer is that it was in the best interest of the Reagan administration for both countries to maintain a stalemate, exhausting its resources and sue for peace. The expected result would be two severely weakened Gulf states that had little will or means to foment revolution in the smaller more vulnerable Gulf monarchies.

If we look deeper into the American selectorate at the time, however, we see that it would have been widely unpopular to supply the Iranians with offensive arms or even more spares for its air force. The outrage over the Iran-Contra affair demonstrated the public's hostilities toward Iran. Indeed, even in 2010 there exists considerable hostility to not only the Iranian regime but Iranians in general. To be sure, sales of more robust weapons would have split the Reagan selectorate and doomed the Republican Party. A move toward appeasement would have been unpopular and undermined the foundations of Reagan's foreign policy. The administration simply endured the criticism, congressional hearings and then moved on to other matters. Halting weapons deliveries did not significantly alter the Iran-Iraq war as by this time both sides had settled into static trench warfare more reminiscent of World War I than the mobile campaigns of Rommel and Patton in World II. The next phase of the low-intensity war between Iran and the United States would center on what became known as the Persian Gulf "Tanker War."

The Iran-Iraq War and the Tanker War

Following an Iraqi attack on Kharg and Larak Island in the Persian Gulf, damaging Iran's oil export capabilities, Iran began attacks on Gulf tankers initiating the opening rounds of the Tanker War (Cordesman & Wagner 1990, Chubin 1989, Farhang 1985, Sick 1989). Recent success on the battlefield made Iran willing to take more chances in order to force an end to the war. At the same time the United States sought to reinforce the Iraqi military positions by providing various forms of intelligence that would allow Iraq to defend against Iranian offensives (Cordesman & Wagner 1990, Chubin 1989, Sick 1989, Stein 1989). Iran sought to gain a strategic advantage and force the Gulf States to abandon their support for Iraq by attacking them at their most vulnerable poin: oil exports. By attacking a source of Iraqi income Iran would not only decrease monetary transfers to Iraq from Gulf oil producers, but would also boost oil prices for the West by increasing shipping insurance which would be passed on to the consumer. Western consumers would, in turn, pressure their government to end their support of Iraq. This two pronged Iranian plan sought to end outside support for Iraq through regional and international pressure. The only drawback was the unforeseen response by the American and other Western governments.

In response to Iranian attacks on tankers of various nationalities the United States initiated Operation Earnest Will on July 24, 1987. The background to Earnest Will was simple. The Tanker War had significantly increased insurance rates for Gulf tankers and thus increased the price of oil. In response, Gulf nations (in particular Kuwait) approached the Untied States Government asking for American naval protection for its tankers. Unknown to the Kuwaitis were statutory restrictions on American naval vessels escorting foreign flagged civilian ships. In order to circumvent the law, the United States and Kuwait arranged for the tankers to be re-flagged as American ships (owned by an American company) despite the company being wholly owned by Kuwait (Crist 2001, Kelley 2007, Pollack 2005, Selby 1997, Stein 1989).

Operation Earnest Will was the American response to this situation. Another reason the American government decided to re-flag the tankers was the incident on 17 May when the United States Navy frigate *USS Stark* was hit by two Iraqi Exocet missiles causing the deaths of 37 seamen and injuring 21 others (Cordesman & Wagner 2003a, Sick 1989, Stein 1988, Talmadge 2008). While Iraq denied they had intentionally targeted an American vessel, it was in Saddam Hussein's interest to increase American naval presence in the Gulf. If Saddam could entangle the Untied States even more it would be just a matter of time before the Americans and Iranians clashed. Iran would lose, thus strengthening the positions of the Gulf States who supported Saddam. With increased support from the Gulf States and a weakened Iran, Saddam would be in a better position for cease-fire negotiations. The incident evoked a strong reaction from the American public who was relatively supportive of the Iraqi's during the war—given antipathy toward Iran over the seizure of the embassy.

A prowling aircraft that is not challenged or illuminated by fire control radars—the standard operating procedure-could easily mistake the warship for a tanker. The crew of the *Stark* did not follow procedures—nor did they take defensive action when they had almost an hour's notice of an unidentified aircraft flying toward their general location. Lack of vigilance on the part of the crew no doubt contributed to the disaster (Cordesman & Wagner 1990, Crist 2001, Kelley 2007, Selby 1997). Whether the Iraqi's were aiming for an American warship or not the result was the same: an increased American presence in the Gulf and a renewed commitment to somehow end the mining of the Gulf specifically, and the Iraq-Iran war in general, were the results.

The re-flagged tankers began a convoy operation on July 24th escorted by American warships. As the convoy steamed through the Gulf the tanker *Bridgeton* stuck a mine (Cordesman & Wagner 1990, Crist 2001, Kelley 2007, Selby 1997). Following the explosion, and containment of the flooding, the convoy proceeded with the wounded *Bridgeton* leading one American frigate (*USS Crommelin*, FFG-37), one destroyer (*USS Kidd*, DDG-993) and one cruiser (USS Fox, CG-33) through the mined waters. Ironically, the *USS Kidd* had been ordered and paid for by the Shah of Iran in 1978. Derived from the Spruance class they were all optimized for Gulf operations. All four ships served in the Persian Gulf much to the consternation of the Iranians. This is but one example of the psychological warfare that has been used by both sides since 1979 in the ongoing conflict. The media image of a wounded tanker leading three American warships demonstrated how vulnerable modern warships were to low-tech, yet highly effective mines. Furthermore, it highlighted how the United States had relatively few resources to find and destroy mines. Admittedly, it can also be said that the Untied States Navy of the late 1980s was designed for the Cold War, deep water operations, not the confined and shallow waters of the Persian Gulf.

On April 14, 1988 an American frigate the *Samuel B. Roberts* struck a mine and was seriously damaged. The *Roberts* was subsequently towed to Dubai on April 16, 1988. While in Dubai the damage was assessed and in May of 1988 the ship was readied for transport back to the United States for repairs. On June 27, 1988 the damaged vessel was loaded on to a heavy lift ship for its voyage. The timing of the *Roberts* voyage is important as its departure, along with other actions, had significant repercussions a few days later (Cordesman & Wagner 1990, Crist 2001, Kelley 2007, Selby 1997). The Roberts incident prompted the American Navy to execute Operation Praying Mantis, an attack on Iranian Navy and oil platforms used by the Revolutionary Guards as bases for the small boats that were harassing Gulf shipping (Cordesman & Wagner 1990, Crist 2001, Kelley 2007, Selby 1997, Sick 1989).

Operation Praying Mantis was the largest confrontation the United States Navy had taken part in since the end of World War II and the largest combined operation Iran had engaged in to that date. The plan for the Americans was to attack Iranian ships while attempting to draw out other ships in order to engage them. Part of the plan was also to neutralize as many of the small boats and bases as possible and cripple the Revolutionary Guard's ability to harass and interdict shipping in the Gulf. In a battle that included a nuclear powered aircraft carrier and nine other vessels, the United States attacked and neutralized the *Saddan* and *Sirri* oil platforms. The Iranian response was to use speedboat attacks on the American ships. These were repulsed with aerial bombardment from carrier-borne aircraft and naval gunfire. When the Iranians deployed the *Joshan,* a fast attack gunboat, it was sunk by American missile fire. The *Wainwright* downed an inbound Iranian F-4, subsequently the other inbound fighters stood-down. The next move came when the Iranian frigate *Sahand* attempted to challenge American ships in international waters close to its base of Bander Abbas. After firing on American attack aircraft, the ship was engaged and hit with at least two Harpoon missiles and four laser-guided bombs. The ship sunk rapidly after its magazine caught fire and exploded. The *Sabalan* attempted to engage the American surface forces but was hit by a laser-guided bomb. Dead in the water, an Iranian tugboat from Bander Abbas was dispatched to take the ship in tow. The Americans decided against pressing the attack and let the damaged frigate be towed back for repairs (Cordesman & Wagner 1990, Crist 2001, Kelley 2007, Selby 1997).

Operation Praying Mantis was a resounding success for American forces, in so far as they destroyed the ability of the Iranian Navy to mount any credible offensive operations against the United States or any of its allies. The operation also established naval superiority in the Gulf that would preclude any further mining operations. American military operations in the Gulf were undertaken by an administration that not only sought an end to the Iran-Iraq war for the sake of the Gulf states but for the free and unimpeded flow of oil from the Gulf. The military action was popular in the United States as the prevailing mood was to strike Iran in any way that was possible. The calculation about potentially ending the Iran-Iraq was secondary or even tertiary to just striking back. The coalition or selectorate that elected President Reagan applauded the military action, as it was not only retribution for the Embassy and the Hostages in Lebanon but also for the destruction of the US Marine barracks in Beirut. The American public had always blamed Iranian complicity with radical Shiite elements for the bombing and the subsequent withdrawal of American forces from Lebanon. In general there was great support from the American selectorate for these and other military actions against Iran in the Gulf.

On the other hand, the Iranian government was concerned about the rising level of tension with the United States. The Iranian selectorate was caught off guard with the "Imposed War" and was not fond of American actions in the Gulf such as helping Iraq and acting as the Saudi's policeman in the Gulf. With the bulk of Iranian naval power destroyed the Iranian leaders could not have been happy, or even willing, to challenge the United States at a game where the US had all the advantages. To keep the social disturbances at a minimum, the most advantageous avenue of action would be to disengage from the tensions with the United States but not Iraq. The American actions did give the Iranian government some important questions to consider, namely would the United States back down or would they escalate the military actions.

Operation Prime Chance

In conjunction with Operation Earnest Will (Aug 1987 to June 1989) the United States tasked its Special Forces Command with deterring Iran from using naval mines to impede Gulf shipping. In a response to losses in its naval forces, Iran began a new strategy in the Tanker War by using various forms of naval mines to slow Gulf shipping. The use of mines goes back to World War I, with mines being a cheap and effective defensive weapon. Moored mines were common in World War II, yet mines do not need to be moored to be effective. A secondary use is strategic area denial. Placement of mines in strategic passages effectively and cheaply denies passage to ships. The effectiveness of an area denial strategy is increased when the party planting the mines has knowledge of the currents and eddies that are present in shipping lanes. By planting mines in currents that will take them into the shipping lanes the planter can lessen their chances of detection and retaliation (Larson et al. 2004, Krepinevich et al. 2003, Truver 2008). Seeding currents could even be done from one's own territorial waters making efforts to deter the mining difficult. In order to disrupt Gulf shipping Iran began to use free-floating mines in the Gulf in mid-1987.

Operation Prime Chance was carried out by American Special Forces, with the goal of interdicting Iranian vessels that were seeding the Gulf with floating mines. While the re-flagging operations of Earnest Will were publicized, Prime Chance was secret (Cordesman & Wagner 1990, Crist 2001, Kelley 2007, Selby 1997). The plan was for Army and lesser numbers of Navy helicopters to interdict the mine laying vessels, which were in many instances revamped fishing vessels or the ubiquitous Arab Dhow. Using night vision devices the helicopter pilots operated from land, ships (Navy helicopters only) and barges (the *Wimblown* and *Hercules*) known as Mobile Sea Bases (MSBs). Attached to these mobile platforms were two or three US Navy Seal patrol boats that followed up on various contacts, boarded suspicious vessels and provided security for the bases. As floating bases the MSBs were moved from one part of the Gulf to another in response to mining activity. The operation lost its secrecy on September 21, 1987 when American forces staged an attack on the *Iran Ajr,* a small ship used by Iranian forces for mine laying. Attacking by air and then by sea the ship was seized, intelligence collected, and then the ship was scuttled. The incident demonstrated that Iran was indeed responsible for mining of the Gulf (Cordesman & Crist 2001, Kelley 2007, Selby 1997, Wagner 1990). Video footage of the *Iran Ajr* laying mines in the Gulf was broadcast worldwide and undermined the Iranian government instance that the mines were planted by Iraq and the United States in order to embarrass Iran.

Operation Prime Chance provided political cover for the re-flagging operation given the mines that were hit in the course of getting the operation going and the lack of mine clearing vessels in the American Navy's inventory. Having to rely upon third parties to clear the mines was an embarrassment to the Reagan Administration and Operation Prime Chance was a way for the Untied States to stop the mining of the Gulf in a secret and effective manner. Politically the success of the Iran Ajr seizure enabled the Administration to demonstrate that Iran was behind the mining of the Gulf. Undercutting the credibility of the Iranian government at this juncture was important for the United States and the Gulf States because both Iraq and Iran were almost to the point where they were ready to call for a cease-fire in their eight year old war.

For Iran the mining of the Gulf was a rational strategy used to intimidate the Gulf States from supporting Iraq. The American response was perhaps stronger than anticipated and the operations were discovered and publicized before the Iranian strategy had a chance to work. In short, the United States had thwarted the strategy. However, it should be noted that the Iranian government had been seeking a way out of the increasingly costly war. While the war was initially popular, as time dragged on and causalities mounted fewer families were willing to allow their sons to join the Basij, whose human wave tactics left scores dead or maimed. There was political will to end the hostilities, yet the tipping point had not been reached at this point in time.

The various military operations undertaken by the United States demonstrated not only how important the United States regarded free navigation of the seaways but it also revealed how it sought to attempt to use what little leverage it had to get the Iranians and Iraqis to accept some type of cease-fire in order to lessen tensions in the Gulf. The last great incident involving the United States and Iran in the evolution of this low-intensity conflict was the shooting down of Iran Air Flight 655 on Sunday July 3, 1988, by the cruiser *USS Vincennes* killing all 290 passengers and crew aboard the Airbus A300B2 (Flight 655) (Cordesman & Wagner 1990, Crist 2001, Kelley 2007, Selby 1997).

The *Vincennes* was providing cover for a heavy lift vessel that was carrying the *Roberts* back to the United States for repairs. Flight 655 took off from Bander Abass bound for Dubai. The flight path skirts the northern part of the Straits of Hormuz, crossing the shipping lanes that lie closer to Iran than the Emirates, and then into Emirates airspace before landing at Dubai. The airport at Bander Abass is a joint military-civilian airfield, which may have been a factor in the American ship determining that the intent of the aircraft was hostile. Early in the morning the *Vincennes'* helicopter received small arms fire from an Iranian gunboat. As the V*incennes* moved to engage the gunboats, they noticed the track of Flight 655 was on a bearing directly toward their position. Flying in a regular commercial air corridor Flight 655 took off then climbed to altitude. The *Vincennes* officers felt that the aircraft could be an F-14 armed with bombs using commercial routes to mask their approach to attack the American cruiser. The warship fired two missiles that destroyed the airliner killing all 290 passengers and crew (Cordesman & Wagner 1990, Crist 2001, Kelley 2007, Selby 1997).

Various factors could have contributed to the tragedy. One factor was the apparent confusion or ignorance of the commercial air routes over the Gulf by the V*incennes* crew. Moreover, American warships had no communications that could monitor normal commercial radio frequencies. The only frequencies they monitored were emergency frequencies (Cordesman & Wagner 1990, Crist 2001, Kelley 2007, Selby 1997). The airliner may have thought that the calls to change course were aimed at an Iranian P-3 *Orion* anti-submarine warfare aircraft operating in the area. The P-3 is capable of firing anti-ship missiles. With the P-3 in the area, the warship was concerned about having not one but two aerial threats while seeking to disengage from the Iranian speedboats. Perhaps the greatest factor at the heart of the tragedy was the sophistication of the Aegis combat system, a computerized system designed to engage aerial, surface, and sub-surface threats automatically and simultaneously. Initially conceived to protect carrier battle groups in open waters, Aegis was out of its element in the confines of the Persian Gulf where hostile, friendly, and neutral aircraft and ships interact. Another contributing factor was lack of training for the crew, who had little experience working with the Aegis system.

The reported aggressiveness of the Captain of the *Vincennes* was a concern. Captain William Rogers was reportedly more aggressive than most captains to the point of actually chasing Iranian speedboats with a billion dollar cruiser not designed for that mission (Cordesman & Wagner 1990, Crist 2001, Kelley 2007, Selby 1997). We do note, however, that Rogers waited until what he thought was a hostile aircraft reached within 15 miles of the *Vincennes* to fire when the rules of engagement called for firing on a hostile aircraft at a range of 20 miles. Given the close quarters of the Gulf, lack of training, the lateness of the flight (27 minutes), other aircraft in the area, unfamiliar computerized combat systems (Dotterway, 1992) and the failure to properly classify the Iranian plane as ascending and the American plane as descending combine to became a deadly mix.

The importance of Flight 655 and Operations Praying Mantis, Earnest Will, and Prime Chance (among others) was that Ayatollah Khomeini reasoned that the game of brinksmanship his government was playing with the United States would eventually lead to an all out American attack. Such an attack would devastate Iran, especially if key defensive installations such as the Silkworm Missile launchers in the Straits of Hormuz or major air and naval bases as are in Bander Abbas were destroyed. Most frightening was the very real possibility that the United States would attack the Kharg or Larrak Islands where Iranian petroleum exporting facilities were located. Loss of revenue would severely undermine the government's finances and perhaps even undermine the revolutionary institutions that had been put in place. On the other hand, the coalition that put Khomeini in power had grown weary of the strict edicts on dress, speech, and travel. The war, while disproportionally shouldered by the lower classes that tended to support the government more vigorously, would have been in danger of losing subsidies and transfer payments that oil revenue provided. Thus, the social bases of the revolution could have been threatened. The more affluent Iranians, who had not seen financial or social gain under the new regime, would have even less reason to support the regime and indeed may have openly opposed the government if the conflict widened or oil revenues were curtailed. To be sure, Khomeini was confronted with losing his main base of support or even risking open rebellion if the United States engaged in open warfare against Iran and Iranian interests.

Given this situation, he decided—in what must have been a galling decision—to seek a cease-fire with Iraq. Perhaps the best way to look at this reversal is the fact that it was the sum total of small defeats that signaled that the United States was serious in its threats and that it would attack Iranian forces at will. The destruction of Flight 655 may have signaled an escalation by the United States that Iran could not counter, save for closing the Straits of Hormuz, which would certainly result in American military action. If the Iranians saw Flight 655 as a deliberate signal that the United States would now engage civilian targets then Khomeini would be correct in seeking an end to hostilities with both Iraq and the United States. This is in fact what he did, thus saving his governing coalition and setting the stage for a redirection of oil revenues to other economic sectors. If the shooting down of the airliner was a mistake as claimed by the American Navy then Khomeini made a practical decision in seeking a cease-fire with Iraq and de-escalation with the United States, given the increasingly severe American military actions as well as the dwindling of Iranian military assets available to confront the United States.

In sum, Khomeini's decision to disengage was made on practical political, economic, and military basis; thus, ensuring that his revolution and institutions would survive intact. Indeed recent evidence suggests that Khomeini concluded from the various American military actions, in particular the destruction of Flight 655 that the United States had decided to undertake unlimited military actions against Iran, given the slightest provocation. Knowing that American forces took great care not to involve civilians, the downing of the airliner must have lead the Iranians to believe that the United States would now target Iranian civilians. Having undergone the "War of the Cities" missile attacks in 1985, the leaders in Tehran were clearly concerned that the United States could mount a much more devastating attacks than had Iraq, and they were concerned that their defenses were not capable of defending Iran. Thus, given months of constant and increasing tension with the United States, which culminated in the destruction of Flight 655, Ayatollah Khomeini concluded that full scale war with the United States was a very real possibility. Thus, he decided that it was time to end the conflict with Iraq and deescalate tensions with the United States (Wilson Center).

The Nuclear Weapons Program

No issue so divides the Islamic Republic of Iran and the United States as the Iranian nuclear program. This is an example of how economic, diplomatic, and military inducements and threats have played out in the past twenty years. This is also an issue that demonstrates how low-intensity conflict goes hand in hand with the theory of the selectorate and how domestic political forces can drive the continuation of a conflict.

This analysis will deliberately focus on the underlying political, economic and military reasons Iran wants to pursue nuclear weapons and why the United States will attempt to stop such a pursuit, rather than the intricacies of the technical details. If Iran were to pursue a nuclear weapons program the calculation would be based on three factors: perceived security threats to Iran; domestic political and economic needs; and national pride or history (Mayer 2004, McNaugher 1990, Minasian 2002, Mokhtari 2005). Iran has always been an avenue for invasion having been invaded by the Mongols, Russia, and having been occupied during World War II. While having a strong notion of greater Iran and its territorial integrity, Iran has not had success in maintaining its sovereignty over what the collective Iranian nation considers the borders of its political state.

The Iran-Iraq War is called the "Imposed War" by the Iranians for the simple reason that it was imposed upon the nation rather than being an option it could choose. Moreover the national sacrifice in terms of foregone economic growth and human suffering was such that the idea of vulnerability pervades most Iranian discourse with the West. The perception of enemies such as Israel or the United States in many respects is political hyperbole aimed at specific groups in Iranian society whose support is needed by the government. Are these threats specious or are they real and immediate? The simple answer begs the question-what does Israel or the Untied States stand to gain by attacking Iran? Clearly the military advantages of the United States are such that a determined campaign could be mounted to invade and replace the Iranian government, yet such a foray would have to overcome a war-weary population who turned against the war in Iraq and where conquering the country was easy as compared to consolidating and calming the population. There is no reason that an invasion of Iran would not turn into an insurgency as happened in Iraq; therefore, what would be the advantage for the United States?

In a similar manner Israel has the military might and will to engage Hezbollah in Lebanon and even Syria in the event of hostilities. Therefore, why would Iran see these nations as security threats? Besides the historical record of invasion, the legacy of the United States, its' support for the Shah, and the revolutionary ideology of Khomeini are the primary reasons for this security phobia. The United States supported the Shah, helped him depose his father and then fomented a coup that swept out the popular albeit eccentric Mohammad Mosaddeq and relied upon severely repressive measures to keep any opposition weak. Thus, the United States is seen as capable of doing almost anything to make sure Iran bends to its will.

In the case of Israel, the issue tends to be more ideological rather than militarily based. The revolutionary ideology of Ayatollah Khomeini is predicated upon the traditional Shi'a understanding of the oppressor and the oppressed. Since the Shi'a have traditionally seen themselves as the oppressed, this ideology seeks to free those who are oppressed from their oppressors. Thus, Khomeini's ideas lent credence to the idea of pursuing war with Iraq. Furthermore, opposition to Israel and support of the Palestinians are extensions of the ideology since Israel "oppresses" the Palestinians who, while not majority Shi'a, are Muslim nevertheless.

The perceived security threat from the United States and Israel, while not grounded in military necessity, does have historical antecedents and can be seen as an extension of the revolutions ideology. In so far as a security threat does exist, such a threat is enhanced by American and Israeli reactions to Iranian actions (as seen in the Tanker War and Lebanon in 2008). The perception of a threat seems to be much more powerful than the actual threat. To be sure both Israeli and the United States could attack Iran easily and cause much damage and human suffering, yet the probability of an attack is extremely low. Clearly this is misperception of a perception, yet as such it is a powerful symbol for the Iranian nation. Is this a rational stance for the Iranian government? The answer lies in domestic and economic dynamics.

The political and economic dynamics of the nuclear power program are interesting and complex. Politically the nuclear program does several things. First, the program mobilizes multiple technical and human assets. The infrastructure has to be built and the technical knowhow to be acquired. Such a program employs people and, in the Keynesian way, stimulates the economy. Second, a nuclear power program independent of a weapons program may actually be in Iran's long-term best interests. Iran holds the second largest reserves of natural gas in the world, yet loses approximately 30 percent of its production to flaring, loss, shrinkage, and gas injection into oil fields. Perhaps the greatest loss is that of 16 percent for enhanced oil recovery of gas injection into oil wells to push the petroleum to the surface. While an effective technique to enhance recovery, the need for this innovation demonstrates three things. First, Iran's oil is running out. Second, the market for liquefied natural gas is not large at this time and third, oil as a commodity brings in considerable revenue, enough so that gas is sacrificed for oil production. These facts play into the political and economic calculations of the leadership on the nuclear power and weapons programs. If declining oil revenues threaten the ruling coalition it is logical that the leaders would seek nuclear power for electrical power generation. It is, however, the dual use of the nuclear reactors that has concerned the West. The type of nuclear reactors being built by Iran (and the uranium enrichment program) point not only to power generation but also to weapons production. This is the primary problem. The political and economic dynamics in Iran are such that it is politically and economically imperative to supplement electrical production as well as provide jobs. A civilian nuclear power program does both. The dual uses, however, are harder to justify economically. The revenue used to build nuclear weapons facilities can be deployed in much more productive areas. Politically, building nuclear weapons can be seen as neutral. Some of the population wants these weapons yet others do not see the necessity. Perhaps another reason Iran seeks nuclear weapons is the third leg of the overall nuclear weapons program, national pride and history.

Iranian history is replete with episodes of invasion, occupation, and counter-invasion. Given such a history it is quite logical that Iran would want some sort of trump card to deter any future aggressor. The idea of national pride is an element of nationalism (Mayer 2004, Mokhtari 2005). Nationalism can take many forms, a benign form being simple patriotism while the most rabid is xenophobic. The nationalism that would support the building of a nuclear stockpile and the associated delivery systems is seemingly aggressive. A closer look at Iranian foreign policy today sees only a few areas where nuclear weapons may be used as leverage to gain an advantage. One would be in defense of its ally Syria and the Hezbollah in Lebanon. However, this ignores the fact that neither one individually or combined pose a threat to Iran's existence.

Iran under the Shah relinquished its historical claim to Bahrain in 1970 in return for other concessions, yet Iran has never given up the psychological notion that Bahrain was ruled for several hundred years by Persians. The perception that Bahrain was going to be ruled by Arabs-even if it was a *fait a compli*-angered many Iranians. While not a threat to Iran, the forcible annexation of Bahrain might be possible if Iran had nuclear capability. The probability of such an action, while perhaps popular in some Iranian circles, would be highly improbable since Bahrain is host to a major American naval base in the Gulf.

The notion of Iran seeking nuclear weapons more for a collective psychological reason cannot be proven empirically. It is certainly possible, however, that joining the nuclear club would give a regime that has been losing credibility with its constituents considerable international respect. Developing, testing, and demonstrating the capability to deliver such weapons would be a boon to the government showing that the Islamic Republic can master the technology required to use the atom. Domestically, such a situation may create civic pride, parades, and a general feeling of support for Iran's new found respect. Yet would a nuclear capability, reverse oil depletion, create employment, reduce urban congestion, or raise the median wage? To be sure the answer is no and as such, any psychological advantage to having nuclear weapons would be short lived and would pose serious problems for the regime internationally. In short, one rational for the nuclear weapons program is national pride, but a deeper analysis demonstrates that national pride is a thin theory to build ones international relations upon.

In the United States there are several reasons for attempting a variety of means to prevent Iran from gaining nuclear weapons and developing the delivery systems that would threaten American Interests. Two non-negotiable American interests are the continued existence of the state of Israel and second is the free and uninterrupted flow of oil from the Persian Gulf. The American government's coalition or selectorate supports both these foreign policy goals and the inability of an administration to ensure both would break up the electing coalition of an administration. Not far behind these two goals are denying Iran from gaining nuclear weapons. While a primary goal, this goal of American foreign policy is not so strong as to break the governing coalition. The recent defeat of Republican presidential candidate John McCain, who stated in no uncertain terms that he would not negotiate with Iran over nuclear weapons and that Iran would be prevented from attaining nuclear weapons, was soundly defeated by a candidate that seeks engagement rather than confrontation with Iran over the nuclear weapons issue. Clearly whether Iran has nuclear weapons is ancillary to the American public at this point in time. However this could change if Iran did explode a test devise that produced results consistent with its design yield. It is possible that the first test might be less than hoped for as India, Pakistan, and North Korea have all experienced less than hoped for results in their initial detonations. Even if the yield of a test weapon is far below Iranian media report, American intelligence services will be able to determine how successful the test is.

While from a technical view the efficiency of the weapon is important, from a political or public relations standpoint it is not critical. What is critical is the public opinion of the weapon test both in Iran and the United States. While Iranian opinion would initially be positive it could turn quickly if the test brought on Western sanctions. Iran is particularly vulnerable when it comes to gasoline, use of international banking facilities for clearing trade payments, and even food. American public opinion would be negative at a moderately high level. While not the primary concern of most Americans, the thought of being in another Middle East war such as Iraq or Afghanistan would quickly force the administration to take some sort of significant diplomatic action to try and contain the new Iranian offensive capabilities. Such actions could be diplomatic, military, or a combination of both. Ultimately the American response would depend on the damage to a President's winning coalition or selectorate that defeat or humiliation would impose. Clearly, once Iran has achieved weapons status the risk increases considerably. The calculation will also be based upon the political party in power in the White House and in Congress. Divided government (one party holding the executive and one the legislative branch) may dilute any response while if one party held both it would increase the chances of a vigorous response.

The amount of risk a President takes is also a function of either being in the first or second term in office. Jimmy Carter was in his last year and a half in office when the hostages were taken. If the rescue mission—Operation Eagle Claw— was successful his political fortunes may have been significantly different. On the other hand George Bush was in his second term when he decided to change strategies in Iraq and engaged in the "Surge" which resulted in breaking the back of the insurgency and hastening the return of the troops. Other considerations went into the calculations of these two presidents but the fact remains, the term and time left in the term has a huge impact on foreign policy decisions. To be sure, an American president would face some tough decisions if Iran "went nuclear' and tested a device. This would be even more serious if Iran possessed a proven delivery system for nuclear warheads.

"Stuxnet" a Continuation of the Conflict

Earlier this year Iranian computer systems were found to be infested with a sophisticated computer virus called "stuxnet." The source of the so called malware is not known but the experts feel that its developers must have had enormous support thereby signaling some state support. Israel and the United States come to mind as nations with the capability and motive to use such a weapon against Iran. The specific virus, while similar to generic viruses is sophisticated in that it only will attack the computer control systems for factories, refineries, and nuclear power plants. As of this date the intended target of the attack is not known with some speculating that the event may have happened. Speculation at first centered around the Bueshehr nuclear power plant that began its fueling in August of 2010. However the intended facility may have been the Natanz uranium refining plant. From a security prospective the Natanz plant is a far better target since it is directly part of the program to derive highly enriched uranium for nuclear weapons. Iran has announced that it has attained the twenty percent level of uranium enrichment considered a milestone. This level of enrichment is not great enough for nuclear weapons but given the physics of uranium enrichment, this batch of enriched uranium can be processed through the cascade of centrifuges only a few more times before it becomes weapons grade,—due to the exponential nature of the reprocessing when a certain (20%) level is reached. Thus, Natanz may have been the actual target. Slowing the nuclear program is of prime importance for Israel and the United States. This first cyber attack on another country, either from state sponsored agencies or from a state supported group, has been launched and may usher in a new age of covert cyber warfare. Assuming that the attack came from the United States or Israel both can expect some type of counter attack. This may come at many levels, from defense department networks to simply hacking and reprogramming the traffic lights in a major city. The former option would be a risky move as defense related networks are well protected but civilian nets are less well protected and simply turning off traffic lights or stopping sewage treatment in a major American city causing social and economic havoc. The major question one needs to address is whether such an attack is sufficient provocation for actual military attacks and what will the selectorate in the United States demand of its leaders. In sum, the low-level conflict has entered a new phase with the battlefield being cyberspace. The vectors of attack are numerous and the damage incalculable and such an escalation could add a whole new level of ferocity to the conflict.

Conclusions

These three episodes of low-intensity conflict between Iran and the United States demonstrate not only how low-intensity wars can be fought but also how domestic forces figure into the equation. The coalition that selects the leadership- the selectorate-has in many instances the ability to drive policy or to restrain governments from escalating to greater levels of violence. In these three episodes of Iran-US interaction the changing nature of the selectorate in each nation, to some extent, dictated the shape of the policy that each nation pursued. In the first episode discussed here, the United States rejected a president and elected a new one who promised strength, while Iran used the War with Iraq and the Hostage crises to rally support for the Revolution and the passage of the new constitution. In the second episode toward the end of the Iran-Iraq war, again, both selectorates were vital in supporting or restraining their respective government's actions. The United States had a selectorate that supported military action against Iran, while the Iranian selectorate only gave weak support for continued military confrontation with the United States. Ayatollah Khomeini's decision to back down and call for a cease-fire with Iraq was prudent given little support for actions that would be a "lost cause" against the powerful American military. The nuclear weapons program is also driven by the politics of the selectorate. While elements within the Iranian leadership see nuclear weapons as a way to enhance their international prestige and fix the institutions as they are now, the United States sees Iranian nuclear weapons as a threat to Middle Eastern stability. Each nation's internal politics will determine how they approach each other and attempt to solve the ongoing low-level conflict.

Bibliography

Bogen, D. & Lynch, M., 1989. Taking Account of the Hostile Native: Plausible Deniability and the Production of Conventional History in the Iran-Contra Hearings. Social Problems, 36(3), 197-224.

Brody, R.A., Shapiro, C.R. & Brody, A., 1989. Policy Failure and Public Support: The Iran-Contra Affair and Public Assessment of President Reagan. Political Behavior, 11(4), 353-369.

Canham-Clyne, J., 1992. Business as Usual: Iran-Contra and the National Security State. World Policy Journal, 9(4), 617-637.

Cavender, G., Jurik, N.C. & Cohen, A.K., 1993. The Baffling Case of the Smoking Gun: The Social Ecology of Political Accounts in the Iran- Contra Affair. Social Problems, 40(2), 152-166.

Chubin, S., 1989. The Last Phase of the Iran: Iraq War: From Stalemate to Ceasefire. Third World Quarterly, 11(2), 1-14.

Cordesman, A. H., & Wagner, A. R. (1990). The Lessons Of Modern War: Volume II: The Iran-Iraq War. Boulder, CO.: Westview.

Crist, D.B., 2001. Joint Special Operations in Support of Earnest Will. Joint Force Quarterly, (Autumn/Winter), 15-22.

Farhang, M., 1985. The Iran-Iraq War: The Feud, the Tragedy, the Spoils. World Policy Journal, 2(4), 659-680.

Hicks, B.D., 1996. Presidential Foreign Policy Prerogative after the Iran-Contra Affair: A Review Essay. Presidential Studies Quarterly, 26(04), 962-977.

Kelley, S. A. (2007). Better Lucky than Good: Operation Earnest Will as Gunboat Diplomacy. Monterey, CA: Naval Postgraduate School.

Koh, H.H., 1988. Why the President (Almost) Always Wins in Foreign Affairs: Lessons of the Iran-Contra Affair. The Yale Law Journal, 97(7), 1255.

Kornbluh, P., 1988. The Iran-Contra Scandal: A Postmortem. World Policy Journal, 5(1), 129-150.

Krepinevich, A., Watts, B. & Work, R., 2003. Meeting the Anti-Access and Area-Denial Challenge. Work.

Larson, E.V. et al., 2004. Assuring Access in Key Strategic Regions: Toward a Long-Term Strategy, Arlington, VA: RAND Corporation.

Mayer, C.C., 2004. National Security to Nationalist Myth: Why Iran Wants Nuclear Weapons, Monterey, CA: Naval Postgraduate School.

McNaugher, T.L., 1990. Ballistic Missiles and Chemical Weapons: The Legacy of the Iran-Iraq War. International Security, 15(2), 5-34. Available at: http://www.jstor.org/stable/2538864?origin=crossref.

Minasian, S., 2002. The Contemporary Status of Iran's Nuclear Missile Programme and Russo-Iranian Relations. Iran & the Caucasus, 6(1), 249-260.

Mokhtari, F., 2005. No One Will Scratch My Back: Iranian Security Perceptions in Historical Context. Middle East Journal, 59(2), 209-229.

Pollack, K.M., 2004. The Persian Puzzle: The Conflict Between Iran and America, New York: Random House.

Posen, B.R. & Evera, S.V., 1983. Defense Policy and the Reagan Administration: Departure from Containment. International Security, 8(1), 3-45

Rubenberg, C.A., 1988. US Policy toward Nicaragua and Iran and the Iran-Contra Affair: Reflections on the Continuity of American Foreign Policy. Third World Quarterly, 10(4), 1467-1504.

Scheffer, D.J., 1987. U.S. Law and the Iran-Contra Affair. The American Journal of International Law, 81(3), 696-723. Available at: http://www.jstor.org/stable/2202027?origin=crossref.

Selby, M.W., 1997. Without Clear Objectives: Operation Earnest Will, Newport, RI: Naval War College.

Sharpe, K.E., 1987. The Real Cause of Irangate. Foreign Policy, (68), 19. Available at: http://www.jstor.org/stable/1148729?origin=crossref.

Sick, G., 1989. Trial by Error: Reflections on the Iran-Iraq War. Middle East Journal, 43(2), 230-245.

Stein, J. G. (1989). The Wrong Strategy in the Right Place: The United States in the Gulf. International Security, 13(3), 142-167.

Sterner, M., 1984. The Iran-Iraq War. Foreign Affairs, 63(1), 128-143.

Talmadge, C., 2008. Closing Time: Assessing the Iranian Threat to the Strait of Hormuz. International Security, 33(1), 82-117. Available at: http://www.mitpressjournals.org/doi/abs/10.1162/isec.2008.33.1.82.

Truver, S.C., 2008. Mines and Underwater IEDS in U.S. Ports and Waterways: Context, Threats, Challenges, and Solutions. Naval War College Review, 61(1), 106-127.

Chapter 8
The Future of the Relationship

The relationship between Iran and the United States has been contentious, and is the first fourth generation war. Low–level conflict is a hallmark of this new type of hostility. The politics of the continued conflict revolve around the various coalitions and domestic actors who assume leadership in their efforts to maintain power, sometimes by force if necessary. This concluding chapter presents several options or scenarios that could lead to improved or normalized relations between the Islamic Republic of Iran and the United States. Most involve the key issue of Iran's nuclear weapons program. This issue is the lynchpin of improved relations. If solved, all ancillary issues such as support for Hamas, Hezbollah, *etc.*, become secondary and eventually resolved. Thus, the resolution of the nuclear issue will settle the low–level conflict that has simmered for thirty years. The dispirit nature of fourth generation warfare makes solving the nuclear issue key to peace and stability in the Persian Gulf and the region in general.

Scenarios for Future Relations

For both sides of the conflict several scenarios stand out as potential ways to either come to terms or to renew the conflict in more violent ways. For the United States the primary calculations of President Obama's Administration revolve around the amount of coercion it applies to Iran and the amount of political capital the President is willing to expend to achieve some sort of settlement to the nuclear, economic, and political issues that separate Iran and the United States. For Iran the calculation is based on the amount of political capital the regime wants to expend to maintain the present shape of its current coalition and the amount of risk the regime wants to take in maintaining its current posture, in particular the nuclear weapons program. There are at least six ways that the stalemate over the nuclear weapons program can be settled from the perspective of this work.[1]

---

1   This chapter borrows terminology and a general orientation from Kenneth Pollack *et al.* 2009, Which Path to Persia? Options for a New American Strategy Toward Iran. Brookings, Washington, D.C. The point of departure taken by this work, however, is that the level of coercion used and the amount of political capital expended on the part of each participant is ultimately determined by the domestic political situation or the selectorate (see Chapter 1). These elements are not discussed by Pollack *et. al.*

Invasion of Iran

The first option is an American invasion of Iran aimed at ending the hostilities; destroying Iran's nuclear weapons program and support for radical groups; and putting in place a new "friendly" regime. While this option does have some merit in that the resources are available for such a venture, the risks outweigh the rewards for the United States. While an invasion causes a high level of coercion, it would also entail the expenditure of a considerable level of political capital on any US president that attempts such a maneuver. Needless to say, the recent US experience in Iraq has tempered the electorate who would not support another Middle East war. On the other hand such a move would ensure that the low–level conflict ended in favor of the United States and that the nuclear weapons program was dismantled. The sense of coercion that this option entails on Iran is exceedingly high from the United States perspective. Clearly, an invasion would be the end of the Islamic Republic, which from the American point of view would be desirable. One aspect of this strategy, short of an actual invasion, that may have merit is the buildup of military forces of sufficient size that the Iranian regime would yield on nuclear weapons and support for radical movements like Hezbollah in Lebanon. The coercive element may be enough to bring the Iranian regime to some sort of accord if they believed an attack was inevitable.

Viewed from the Iranian standpoint, significant pressure by the United States in the form of an invasion would place considerable pressure on the ruling coalition of hard liners, the radical clerical establishment, and the Revolutionary Guards to either prepare a defense or yield to American demands. If the regime gave in to US demands it would expend its political capital in such a manner that it would lose its legitimacy and be an easy target for reformers to foment a Velvet Revolution. Alternatively while the military and Pasdaran (Revolutionary Guards) are fully aware that they could not resist the American military, they could make the invasion costly. The calculation of risk the regime could tolerate would ultimately be based upon the general population, who at least at the initial stages of an invasion rally around the flag and support the government. Even the buildup of American forces would reinforce the idea in the overall population that the United States wanted to invade Iran for its own purposes rather than eliminating an unpopular regime. In either instance the calculation for the Islamic Republic's leaders would be to confront the US and dare them to invade. This sort of brinksmanship is inherently dangerous for both sides.

Ultimately this strategy imposes too much risk on Iran (that is the fall of the government) and would necessitate a huge expenditure of political capital on the part of the US administration. While the possibility of a Velvet Revolution would exist given a possible American invasion, there is no guarantee that a government brought into the fore by fear of an invasion would be on good terms with the United States. An ultra–nationalist Iranian government might be as difficult to deal with as the Islamic Republic. An American administration would pay a high political price in elections if an invasion was long and drawn out or brinksmanship resulted in another hostile regime. Given these circumstances it is highly doubtful that the Islamic Republic would have to withstand an American invasion or the United States would seek this sort of solution to the conflict. In sum, this scenario is exceedingly unlikely.

American Airstrikes

A second option for the United States would be airstrikes aimed at destroying the Iranian nuclear weapons program and some of the infrastructure used to support radical groups like Hezbollah. The US calculation is based upon two variables: the expenditure of political capital and the level of coercion the administration would be willing to apply. The Iranian calculation is based upon the variables of political capital that the regime must expend to hold its coalition together, either through incentive or force, and the level of risk it is prepared to take in defense of its coalition and position.

Air strikes on Iranian nuclear weapons facilities involve a relatively high level of US coercion while entailing the expenditure of relatively little political capital. The US clearly has the military resources, the bases, and the political will to conduct large scale airstrikes, but this would not be the best strategy if the goal is to end decades of low–intensity conflict. Clearly the objectives of destroying the nuclear weapons infrastructure will be accomplished, but the residual effects of such an attack would be significant.

The Iranian calculation on how to respond to an American attack again revolves around the amount of political capital the government wants to expend to keep power and the risks associated in absorbing the airstrikes or lashing out in other venues. Politically the government will gain significant capital if the US strikes and the Iranian government calls for some sort of retaliation. No matter what will be forthcoming from the more radical segments of the ruling coalition, even moderate groups will accept some sort of retaliation for the violation of Iranian sovereignty. Given the historical interventions by the West, any attack would not be accepted in Iran even by those who oppose the government. The most substantive calculation the Iranian government must make is how much risk it is willing to accept on its coalition if it uses various means to retaliate against the United States.

Potential avenues of retaliation include activities in the Persian Gulf such as mining, use of missile batteries in the Straits of Hormuz, and missile boat attacks on tankers or warships in the Persian Gulf or Straits of Hormuz. Perhaps most important would be the rise in petroleum prices prior to any attack and especially after an attack. Not only would the price of oil itself increase dramatically but insurance rates for tankers entering the Persian Gulf would be exorbitant and would be passed through to consumers.

An alternative strategy would be for Iran to use Hezbollah in Lebanon to attack Israel with rockets or to actually invade Israel using specially trained commando units. If Israel were infiltrated under the cover of a rocket bombardment, these units could occupy Israeli–Arab villages, impeding the Israeli Defense Force (IDF) from moving units to the Lebanese border. More important would be the fact that Israel would have to engage in slow, costly urban combat. Combined with hostile moves by Syria this would make for an explosive scenario. Surprisingly this strategy would have little risk for Iran unless the US allowed Israeli over–flights above Iraq to strike Iranian targets. What Israel would strike would be an interesting question if the entire nuclear weapons infrastructure was destroyed by the US.

An alternative would be economic targets like refineries or the main shipping terminal at Kharg Island in the Persian Gulf. Such an attack on economic targets could also increase the price of oil and damage the global economy. Bringing in proxies would be a relatively low risk strategy for Iran but could have considerable consequences both politically and militarily, especially if Hezbollah is destroyed militarily. While this scenario is more likely than an invasion, it has several disadvantages especially in provoking Iran into using its proxies to cause problems for Israel or perhaps even the global economy. The probability of this sort of attack is directly associated with the American perception of Iranian willingness to negotiate over the nuclear weapons program.

Israeli Air Strikes

The third policy option available to the US and Iran would be to do nothing and let Israel conduct airstrikes and destroy the Iranian nuclear weapons facilities. From the US perspective this would involve relatively little political capital expenditure and relatively high coercion aimed at Iran. From the Iranian perspective if the United States tacitly allowed a strike, or even sanctioned it, Iran could face a high level of coercion but the government would actually gain political capital.

From the US perspective letting the Israelis do the dirty work has obvious advantages. First, the administration looses little political capital in the initial stages but could lose considerable capital if the United States had to intervene in some manner. Political capital could be lost in two different ways. Internationally, if the US allowed over–flights above Iraq, the situation in Iraq could become tenser at precisely the time when American troops were leaving. This could allow for a renewed insurgency. Alternatively, should the government of Turkey allow over flights it could encounter internal opposition. Iran could also cause problems for US forces in Afghanistan if an Israeli strike was sanctioned by the US. The resulting fallout from an Israeli strike would be that the US would look complicit in allowing the attacks and would lose credibility in the Middle East.

For the US the most dangerous outcome of an Israeli strike is an attack upon Israel by Iran or one of Iran's proxies, namely Syria or Hezbollah. In this scenario Iran's risks are high but a relatively low expenditure of political capital is probable if (and until) Iran decided to retaliate. After an initial Israeli strike, the Iranian government coalition would be strengthened by further popular support.

Depending on the type and magnitude of retaliation, however, support could evaporate and even threaten the coalition. The most likely retaliation would be a massive missile barrage and commando style invasion of northern Israel by

Hezbollah forces. This would be politically advantageous for the coalition in Tehran as it involves little risk to Iran itself. If Israel deployed the full weight of its military, however, Hezbollah would be decimated and southern Lebanon ravaged by fighting. An uprising by Hamas in Gaza would give a pretext to re–occupy parts of Gaza and destroy Hamas militants.

The destruction of either group would most likely put severe pressure on the governing coalition in Iran to support the client groups in some manner. Upping the ante by Iran launching a barrage of missiles from Iranian territory into Israel would solidify the coalition yet perhaps not garner further popular support based on the popular opinion that if struck, Israel would retaliate. A small number of missiles could be intercepted and destroyed but a large volley fired simultaneously could overwhelm Israeli defenses and inflict some damage on Israel. The calculation for the Iranian leaders involves how much they are willing to risk for the sake of retaliation. While some damage to Iran's proxy groups would be a propaganda booster for the government a potential counterattack could fracture the ruling coalition with demands for increased lethality in a follow up attack on Israel. Thus, the government could be faced with a situation where to hold the coalition together, and hold public support; they may resort to chemical or biological warfare. Such a move entails significant risk. Israeli retaliation would be swift and devastating. Most likely nuclear weapons would be launched from missiles, airstrikes would target major military targets, and population centers could even be attacked. This is indeed the nightmare scenario and as such the Iranian government would have to measure its response to an Israeli attack very carefully.

The least risky option would be a partial or even temporary closure of the Straits of Hormuz thereby increasing oil prices in order to force the United States and Europe to halt initial or further Israeli retaliatory strikes. Clearly the Iranian governing coalition would find the need to retaliate in some way. The question is how much risk they would be willing to take. This risk would be both military and political. A miscalculation could bring further retaliation or even a crumbling of the Iranian coalition due to popular unrest brought on by a population that would blame the government for retaliatory strikes. The population may even face some sort of increased economic sanctions that the West would impose following an Iranian attack, or a closure of, the Gulf to tanker shipping.

This is one of the most dangerous scenarios for all three nations. The only nation whose governing coalition would actually be strengthened, even with retaliatory strikes, would be Israel. A US administration could lose political capital if it was perceived as letting Israel do its job for it. Conversely the United States could lose significant Arab support if it supported Israeli actions, despite the fact that many Arab nations (and Turkey) would welcome either US or Israeli actions against Tehran's weapons program. The Iranian governing coalition would stand the most to lose if its weapons program was fully and completely destroyed. While the probability of total destruction is relatively low, significant damage setting it back five to ten years would have a significant impact on the ruling coalition. Low risk retaliation would encourage the general population to rally around the government but would have little lasting effect on the government's popularity. Therefore, the government of Iran is in an unenviable position given the structure of the governing selectorate. In this scenario all three nations could lose substantially if missteps lead to retaliatory cycles and escalation.

Persuasion and Engagement

An attempt could be made by the United States in conjunction with the EU, Russia, and China to "persuade" Iran to abandon its hostility toward the West and to halt its nuclear weapons program. For either the persuasion or engagement option to be effective it would have to be applied by all interested parties. This is the major weakness of such a strategy for ending the conflict between Iran and the US (and by extension, the West). The EU is fractured, Russia has a different set of priorities, and China (wanting to ensure access to petroleum) would veto any truly meaningful sanctions. Thus, this would be a tricky strategy for the US to attempt. The expenditure of political capital and potential destabilization of the US governing coalition could occur if persuasion and engagement did not halt the nuclear weapons program.

The goal of persuasion is to persuade, through various incentives and coercive measures, to halt the nuclear weapons program and stop the support of radical elements in the Middle East. Conversely, engagement entails a talking approach without the coercive measures that persuasion includes. This can be done either incrementally or through a "Grand Bargain" that puts all the eggs in one basket in a take it or leave it deal.

Persuasion involves both incentives to direct Iranian actions toward desired US goals and some sort of sanctions if Iran does not meet those goals. Essentially this is a carrot–and–stick approach from the US perspective and is a desirable strategy since it involves incentives and relies upon Iranian actions to determine the next move. Conversely, from the Iranian point of view, this is simply another form of colonialism where Iran is threatened and forced into an unequal set of requirements that is neither proper nor just. From the Iranian standpoint, being part of a persuasion strategy would entail relatively low risk as long as Iranian actions were viewed as moving in the direction favored by the United States. However, domestically, the ruling coalition could not be seen as being directed nor controlled by the West. Thus, real progress would be hard to make by simply talking to the Iranian government. Using incentives would work much better but could call the legitimacy of the regime into question and thus be less effective than hoped for. The use of some sort of economic or financial sanctions might bring the regime to the table, but coercive measures would cause the Iranian government to retrench rather than see an end to economic and financial sanctions. Targeted sanctions may not work as well as the West might hope. For example, the most talked about sanction (gasoline) would not hurt those in power or the powerful coalition that makes up the selectorate of the government. The common person and the family would be the ones harmed by an end to gasoline imports. Rising fuel costs would hurt consumers and the middle class the hardest. Most of the resentment would be directed towards the West (the United States in particular) since sanctions of this type would be seen as coming not from missteps of the Iranian leadership but from the outside (something the leadership has little control over). Persuasion may work to some extent but it may not produce the results the United States wants.

From the American perspective persuasion entails little in the way of coercive risk to American or Allied military forces. However, political capital could be quickly depleted if Iranian behavior remains static. While the American public has been at odds with Iran since the seizure of the American Embassy there seems to be "Iran fatigue" when it comes to the carrot–and–stick approach, especially the nuclear weapons program, to improve the relations. The US administration would risk political capital if it made significant concessions only to be rebuffed or not receive an equally tangible concession in return. Moreover, canny diplomacy could prolong the talks for years, allowing Iran to complete its nuclear weapons program or test and produce long–range offensive armaments. Such possibilities would drain political capital and make such a strategy for ending Iran–United States hostilities a multi–administration endeavor. Given the idiosyncrasies of US politics, it is doubtful that such a policy could be sustained for the years needed to gain the trust and negotiate the tough details to the satisfaction of both parties.

The idea of engagement with Iran is an alternative to the strategy of persuasion. In essence this is simply a strategy of talking. From the American position engagement would entail the "Grand Bargain", meaning that diplomatic relations would be resumed, the nuclear weapons program would be halted, support for Syria and Hezbollah would be discontinued (as would support for any other radical organizations). From the Iranian perspective the United States would have to apologize for any aggression, such as to the coup against Mosaddeq, support for the Shah, etc. Economically the United States would have to reverse any and all sanctions and allow exports of food; basic materials; and high tech equipment for Iran's 1970s vintage, US built refineries and oil production equipment. Allowing Iranian assets to be unfrozen would also be a condition.

Difficulties arise when it comes to support for radical groups and the nuclear weapons program. An accord not taking these two issues into account would fall short of its goals and eventually fall apart. Would Iran end support for Hezbollah? The current makeup of the government of Iran would have a very difficult time withdrawing support from their Shi'a compatriots in Lebanon. Politically, this would cost them significant political capital and thus be risky for the regime. To withdraw support the Iranian regime would need guarantees and incentives that are greater than simply goodwill. A security guarantee that Israel would not strike Southern Lebanon may be an option, as would an autonomous province inside Lebanon for the Shi'a run by Hezbollah. Moreover, Iran would probably want some sort of non–aggression treaty with the United States to ensure that they would not be attacked.

While these are certainly possible within the larger framework of a "Grand Bargain," the nuclear question is a much thornier issue. Since the Iranian government and the coalition that supports the government has staked much of its legitimacy on the protection of Iran, and the establishment of greater Iranian power in the region, backing off the nuclear program would prove to be very difficult indeed. The loss of political capital without some sort of *quid pro quo* might be enough to splinter the current governing coalition and throw the nation into chaos. Thus, some sort of tangible *quid pro quo* would be necessary for the government to sanction this sort of arrangement. First, nuclear power reactors would have to be supplied to Iran in sufficient quantity to meet its energy needs for the next 20 years. These reactors can be thorium based which are incapable of producing weapons grade fissile material. Second, some sort of security guarantee would be needed that ensures the sovereignty and integrity of the Islamic Republic's government. Some sort of non–aggression pact between the US, NATO, the EU, Israel, and the GCC would be needed. Furthermore, Iran would need to field a military that would be of sufficient strength to discourage any attack. A modernization of the Iranian forces would be necessary perhaps with French or British equipment. Third, OPEC may find a formula that would allow Iran to increase its oil production to boost revenues, as well as Western assistance in developing its vast gas reserves to increase the lucrative liquefied natural gas export trade.

The above begs the question: 'Would the West and Iran be able to live within the framework of a Grand Bargain?' To be sure, the West would have to offer much more than it would receive in return. The offer would have to be such that it would not undermine the legitimacy of the regime, while also being an offer that would be hard to refuse. On the other hand, the United States government would be put in a vulnerable position if it gave too much and did not receive tangible returns for its investment and efforts.

Containment

In the Cold War the United States used the Doctrine of Containment to "contain" Soviet expansion and power. Such a move by the United States toward Iran would follow the same general pattern yet the result could be much different. Containment of Iran simply means that Iran would be contained or prevented from interfering with US interests or Allies in the Middle East. The implicit assumption of this strategy for ending the conflict with Iran is that it assumes Iran will develop, test, and deploy nuclear weapons capable of striking Israel, Russia, and Europe.

For the United States, a policy of containment would be different from the Cold War era containment. First, Iran will not be a superpower; yet, having leverage over the Straits of Hormuz gives Iran potential power beyond its foreseeable nuclear arsenal. While the United States and the Soviet Union had allies, surrogates, and proxies during the Cold War, both Iran and the United States would as well, however, Iran's would be more contained and on a smaller scale than were those of the Soviet Union. As a regional nuclear power Iran could wield considerable influence by threatening Israel, Saudi Arabia, and the United Arab Emirates. Following United States withdrawal from Iraq, Iran could flex its muscles by attempting to invade and reintegrate Bahrain into mainland Iran. Would the United States risk its military to save a small island nation? The calculation would be based upon the American ability to deter Iran from striking Israel, Saudi Arabia, and other American allies in the region (or even Europe). Clearly it would be difficult for the United States to extend its deterrent umbrella over friendly states in the entire region. Therefore, nations who are threatened by a nuclear Iran such as Saudi Arabia, Egypt, and Turkey may try and attain nuclear weapons themselves to balance and create a stable, regional nuclear deterrent.

One problem with the idea of a regional deterrent (minus Israel) is the stability of constituent governments. Hosni Mubarak of Egypt is aged and thus far has not named a successor. While the succession of King Abdullah in Jordan was smooth, the much larger and diverse social and economic actors in Egypt may make a smooth transition much more difficult. A stabile government may not be possible for years. Saudi Arabia possesses similar problems with its line of succession, the lack of democratic institutions, as well as traditional hostility toward Israel.

The state of Israel would need ironclad guarantees for its security from an Iranian nuclear arsenal. While Israel could defend itself very adequately against an Iranian nuclear attack, its small size makes it exceedingly vulnerable to a single nuclear strike or even near misses. Anti–missile defenses, while impressive, can be overcome with barrages of missiles and decoys. Massive volleys of short–range rockets fired from Lebanon could saturate or cloud the defenses and allow at least one warhead to get through. American assistance would be helpful but integration of the two defensive systems could be difficult and allow gaps in coverage. Simply put, Israel cannot physically withstand one nuclear strike and remain a viable state. Given this reality it is rational for Israel to adhere to a doctrine of launch on warning if they confirm an incoming Iranian missile—no matter if it is conventional or nuclear tipped.

From an existential point of view, Israeli decision makers have little choice to adopt an alternative doctrine. Such shaky deterrence is inherently destabilizing and could easily lead to an accidental nuclear exchange. Both Iran and Israel would have an incentive to either launch or lose their arsenals, which is perhaps the most destabilizing strategy in a deterrent situation. Containment without some sort of mechanism to create stable deterrence, or in some way to assure Israel of its survival, would not be a containment that was strong enough to deter.

One issue not mentioned in much of the literature on the Iranian nuclear program revolves around a possible Iranian nuclear doctrine. While it is beyond the scope of this work to examine such a doctrine, a nuclear power must conduct activities such as building some sort of target list and developing procedures for command and control. India and Pakistan, with Western assistance, have developed protocols that would prevent the unintended usage of their nuclear arsenal. The possession of nuclear weapons entails an aspect seldom discussed—rationality. Will a nuclear Iran be a rational state? The answer will certainly be yes despite the harsh rhetoric from Tehran. Thus far all nuclear powers have demonstrated the utmost in rationality in their dealings with one another. While India and

Pakistan are serious regional rivals and have fought several wars they maintain a stable deterrence despite the vitriol. Combine this with the American and Soviet animosity during the Cold War and we have precedents that show how deterrence can be practiced and that deterrence is a rational undertaking. Tehran has been rational in its international relations and it can be expected to be rational if it crosses the threshold and becomes a nuclear power.

The primary reason for a rational system of deterrence is the selectorate or ruling coalition. An Iranian strike on Israel would initiate an Israeli response many magnitudes greater than the damage inflicted on Israel. Would the ruling Iranian coalition risk losing power, the national infrastructure, civilian casualties, and revenue in a futile spasmodic nuclear strike on Israel? The answer is doubtful. Thus, it is safe to conclude that in the same manner as the Soviet Union exercised nuclear restraint and exhibited rationality after building a nuclear arsenal that Iran would do the same.

An even more appropriate example would be the Peoples Republic of China. In a similar manner China built and tested nuclear weapons, while also threatening the existence of the Nationalist government on the island of Formosa (or Taiwan). The prime deterrent to the Peoples Republic was US support for the Republic of China and the Nationalist Party. This situation has remained stable since the 1950s and should remain so in the future. In the past twenty to thirty years, considerable trade between the two Chinas has increased exponentially. While no diplomatic relations or flights exist between the two nations, families have been reunited and considerable commercial transactions occur each year. This relative stability has been strained at times, but overall there is a stable deterrent system in place. Perhaps the same deterrent could be developed between Iran and Israel. If enmity was restricted to Iran and Israel, stable deterrence could be achieved. Given the multiple interests of various other actors such as Egypt, Turkey, and Saudi Arabia such a system would be far less stable than nuclear deterrence has been in the past.

One way in which a nuclear Iran could be deterred would be some sort of settlement of the Palestinian question and the incorporation of Israel into the NATO alliance. This would solve one of the major problems most Arab states have with the Israelis, while at the same time provide protection above and beyond what the United States alone could provide. Moreover, security arrangements between the GCC and NATO would extend the security umbrella to the Gulf and avert a potential nuclear weapons race in the Gulf, Turkey, and Egypt. Extending NATO's umbrella of deterrence to Israel and providing security guarantees to the GCC would be a considerable accomplishment for any American president or even the majority of NATO leaders. Such an arrangement would be one way to contain the spread of nuclear weapons and provide stable deterrence between the Islamic Republic of Iran and Israel. While this framework for security may never evolve it would provide a workable security structure for the entire region.

Regime Change

One way to change Iranian relations with the US would be a regime change. Regime change could take three forms: insurgent groups eventually take control of large swaths of the country, the military stages a coup, or a broadly based Velvet Revolution much like the one that toppled the Shah.

From the US standpoint it would be advantageous for an insurgency by Arab Iranians or Baluchi's to destabilize and finally overthrow the current regime. Yet it is highly unlikely that the US or any other combination of Allied nations could bring about such a result. The cost and logistical support necessary would no doubt be possible but a US administration devoted to this strategy could be easily dissuaded by insurgent defeats, funding suspensions (as happened with the 'Contra's'), or simply the loss of resolve. On the other hand, the Iranian government could easily take care of multiple insurgencies as it has done in past years. Moreover, the nature of society in Iran itself is such that hundreds of years of invasion and assimilation have created an ethnic amalgam that in some ways resembles the United States. Most people are an ethnic mix and identify themselves as Iranians or Persians rather than Azeri's, Kurds, or Arabs. This situation alone would make an insurgency almost impossible to start, sustain, or prevail over the current government.

The potential for a military coup to displace the current government is only slightly higher than a strategy of insurgency. First, while a relatively low coercion strategy, the US administration could lose considerable political capital if the coup failed. Furthermore, if the coup plotters were successful yet were seen as puppets of the United States, the new government would have little legitimacy and thus fall relatively quickly. Alternatively, if a coup plot was discovered and tied to the United States, dissident elements would be discredited and any associated figures would be jailed or even executed. Moreover, the stigma of American meddling in Iranian affairs would strengthen the hand of the government in dealing with all dissidents.

From the Iranian viewpoint a coup attempt would prove that the United States was hostile and set relations back. The thought of another coup like the one that brought the Shah to power after the Mossedeq interval would sour relations with not only the United States, but the entire West. Considering the above scenario, fomenting a coup against the Islamic Republic would not possess a high enough probability of success to warrant investigating the possibility. Given the slight possibility of success, this would make such a strategy extremely risky for any US administration. Simply put, the risks outweigh the rewards for the United States. For Iran, an American inspired coup would be easily preventable given the dual nature of the Iranian military apparatus (the Revolutionary Guards and the regular military) as it would be difficult for one to plot and carry out a coup without the support of the other. Given Iran's internal security apparatus it is doubtful that a meaningful cadre of coup plotters could be assembled and the necessary logistical and communication networks be assembled without drawing attention to the activities. The probabilities of an American sponsored coup or even an indigenous coup plot are exceedingly small at this time.

Velvet Revolution

Lastly, the one possibility for a solution to the long–standing state of conflict between the United States and Iran would be a "Velvet Revolution." Regime change would be one way Iran could maintain its sovereignty yet have a legitimate government given such a change would come from within the country, not with Western prodding. While regime change would be welcome in the West, any overt support for opposition forces could backfire and allow the government to claim outside interference and use repressive measures to defuse the situation.

The dynamics of a Velvet Revolution are simple. When an authoritarian government becomes repressive, steals an election, or engages in delegitimizing actions it will lose its monopoly on the use of force and the population will take to the streets, participate in strikes, and in general shut down the nation. While the first so–called Velvet Revolution occurred in Czechoslovakia, there have been others, yet the same general dynamic remains. First, an authoritarian government makes a mistake and perhaps even inadvertently kills demonstrators, thereby loosing what little legitimacy it had. The fallen demonstrators become a rallying point for the whole populace who has grievances against the government. Second, people take to the streets and participate in strikes that paralyze not only the economy, but the government in general. Third, the failure to respond with massive forces (because security forces now back the demonstrators) causes the government to lose all legitimacy. The government is forced to negotiate for a transition to new leadership.

While Velvet Revolutions tend to be far less bloody than social revolutions (French, Russian, Chinese, and Iranian) they can be stopped by the government if the government maintains the loyalty of key units of the security forces. As we have seen in the aftermath of the recent Iranian elections, the government will maintain order. The structure of the Iranian security forces makes this possible. In addition to military forces of the Revolutionary Guards (the Pasdaran), and the regular military, there are local para–military volunteers or Basij. Either the Pasdaran or the Basij could be used to calm protests and defuse any potential Velvet Revolution. As recent events have shown, these para–military forces are effective in quelling protests and will be more so in the future. The summer 2009 demonstrations included the first widespread use of mobile communications devices to help network the demonstrators. While networking devices can certainly help protesters avoid authorities, their use also gave authorities avenues to infiltrate, locate, and eavesdrop on protesting groups. As was seen, the government is keen on not letting protests undermine the legitimacy of its rule.

While a Velvet Revolution would be the best possible outcome for the United States, from the point of view of the Iranian government it is an event that must be avoided. Naturally all governments, and in particular the coalition of interests that keep them in power, do not want to see their downfall. Thus they will use almost all means at their disposal to ensure their survival. This would seem to work against the odds of a Velvet Revolution overthrowing the institutions and leaders of the Islamic Republic. The probability of a Velvet Revolution happening in Iran at this time is almost zero and will continue to have a low probability until such time as the government becomes as repressive as the Shah's regime, the vast majority of the populace revolts, and governmental legitimacy is destroyed. Given the dispirit power centers that constitute the selectorate or ruling coalition in Iran it seems highly improbable that one of the major coalition members would pull out and create a vacuum in the power structure.

The low level conflict between the United States and Iran has little chance of ending in the near future. Both nations find it advantageous to end the conflict but as this work has noted the domestic political conditions direct policy makers more toward confrontation than cooperation. In Iran the June 2009 elections and the aftermath demonstrated that the country sees its survival as paramount. Democracy is used to give legitimacy to the candidates that the government chose to put on the ballot. Americans see this system as illegitimate and corrupt. As such the American public has little sympathy with the Iranian government but supports the population at large. The paradox for the American popular opinion is that while supporting the "people" of Iran in their quest for true democracy, there is little patience with the government and its endeavors. In the United States policy makers are intent on stopping the Iranian nuclear weapons program. Conversely in Iran the government steadfastly defends its policy of peaceful nuclear development. Each side has marshaled allies to its side. For the United States the Europeans for the most part have assisted with economic sanctions, but are hesitant to lose the income many firms garner from Iranian operations. Iran counts China and Russia as tacit allies in its quest for nuclear power. Both Russia and China have ulterior motives in assisting Iran, primarily they seek to confront the United States. Russia seeks to make itself an energy superpower and regional hegemony. The invasion of Georgia and the recent treaty with Ukraine over the Crimea point to Russian expansion in the region. Chinese, help in expanding Iranian gasoline production and exports of consumer goods point to Chinese efforts to confound American interest in weakening Iran in order to force an end to the nuclear weapons program.

Another area where Iran and the Untied States continue to battle in the low-level conflict is the cases of Lebanon, Iraq, and Afghanistan. In Lebanon Iranian support for Hezbollah has increased and threatens the delicate political balance achieved at the end of Lebanon's civil war in the 1980's. Furthermore, Israel has security concerns over the military power of Hezbollah in southern Lebanon. If in a conflict, Hezbollah were to open an additional front with Israel, the Israeli's would have to devote resources to meet the threat. Massive short and medium range missile attacks by Hezbollah could disrupt Israeli mobilization and logistics not to mention cause civilian causalities. Such a situation would not go unpunished by the Israeli's who would bring the war to Lebanon in a manner that forces the central government to reign in or engage Hezbollah in order to limit their ability to strike Israel, or to actively engage in a war with Israel. This area is one where the potential for the low-level conflict between the United States and Iran to evolve into a larger more active shooting conflict.

The bellicose tone taken by Iranian leaders toward the United States works against the American administrations attempts to engage Iran. Recent remarks by the Iranian President only hours after the American President extended the offer of negotiations and engagement on the nuclear and other issues will only prolong the low-level conflict. Such remarks allow the American selectorate to solidify its support for candidates who want to take a harder line against Iran—including those who support military options. A more nuanced answer would have disarmed the American right wing while buying time for Tehran to test a nuclear weapon or at least ward off new sanctions against the regime. This tactical blunder was not a wise move considering past Iranian diplomacy which has been nuanced and sophisticated.

The recent revelation about the "stuxnet" computer virus and its potential to do harm to Iranian civilian and military installations is a serious escalation of the low intensity conflict between the two nations. The target is not known and if the virus did indeed attack a facility its target is not known. For obvious reasons the Iranian government would face domestic problems if they were seen as weak or impotent in the face of an Israeli or American cyber attack.

Final Thoughts and Conclusions

This work has presented the relationship between the United States and the Islamic Republic of Iran as a low–intensity conflict, based upon notions of a new type of conflict—the fourth generation war. Wars are carried out by states and each state has a different institutional, economic, and social structure. These structures can be deconstructed into various interest groups of varying size and influence. The size of the winning coalition can determine what foreign policy moves are made and how much risk a government is prepared to take to advance its agenda. In many instances it can be seen as a trade–off between the expenditure of political capital and risk, or how much coercion one intends to apply.

In the case of Iran and the United States, the cost of continuing this low–level conflict is relatively low, which gives some insight as to its prolongation. Each side has vilified the other and governmental leaders can easily call Iran or the United States a villain, justifying their policies as a response to the others actual or perceived intentions and actions. This strategy by both sides has been fruitful in holding the winning selectors of each respective nation together for over thirty years. Has this strategy run its course? With the advent of a viable Iranian nuclear program capable of building, testing, and delivering a nuclear weapon anywhere in the region within the next five years, the low–level conflict will cease to exist and a Middle Eastern Cold War will become a reality.

The relationship between Iran and the United States resembles two young boys. The smaller seeks protection from neighboring bullies, while the larger not only provides protection but also enjoys the friendship of the smaller boy. A deep friendship evolves, because they are inherently similar on multiple levels, but when the younger, smaller boy grows and does not need protection, the protector does not understand and forces his views on the younger friend. The ensuing revolt against demands, not seen as friendship but manipulation, is not welcome. As happens so often, words are exchanged followed by blows and each sulks off vowing never to speak to the other again. However, as young boys return to rational thinking a new status quo evolves, new boundaries are established, and they realize that they not only like one another but they need one another. Eventually the friendship is renewed. Perhaps with time and diplomacy the low–level conflict between Iran and the United States can be settled amicably and justly so the friendship can return and grow.

Index